Potential Theory Applied to Cathodic Protection Design

Rogelio De Las Casas and Ronald Bianchetti

AMPP™

AMPP
Association for Materials Protection and Performance

Printed in the United States of America

Print ISBN: 978-1-57590-495-5
eBook ISBN: 978-1-57590-496-2

All images courtesy of the authors unless specified otherwise.

AMPP
Association for Materials Protection and Performance
15835 Park Ten Place
Houston, TX 77084
ampp.org

Contents

Preface

Potential Theory Applied to Cathodic Protection Design provides the mathematical development of analytical equations for the design of cathodic protection (CP) systems for underground structures with anode groundbeds having various configurations. Equations include calculations for the resistance and voltage (potential) and will assist the CP design engineer.

All the previous references on this subject matter (including NACE CP courses) utilize equations developed almost a century ago by Sunde and Dwight for electrical grounding systems. This book also has a methodological approach, showing detailed developments and practical examples to the equations.

The book is unique in the presentation of the theory that justifies the development of equations to design and simulate CP systems. The book provides the methodology to develop new equations with more specific uses, if required.

- Step-by-step development of the original equations by Sunde and Dwight.
- Explains the development of new equations for the design and simulation of CP systems.
- Explains a new approach to the categorization of DCVG anomalies, based on potential theory.
- For the first time, the authors explain the correlation between the equations and actual field testing, and provide the procedures for field verifications.
- Explains the CP attenuation on transmission pipelines to include the influence of the groundbed and the influence of neighboring pipelines in the same right of way with different attenuation characteristics.

The authors want to thank Elsyca for their assistance in verifying the mathematical developments and calculations presented in this book.

Equation	What is calculated?	Equation number
$V(x,y) = \frac{\rho}{4\cdot\pi}\cdot\frac{2\cdot I(0)}{L}\cdot\left[\ln\left[\frac{x+\frac{L}{2}+\sqrt{y^2+\left(x+\frac{L}{2}\right)^2}}{x-\frac{L}{2}+\sqrt{y^2+\left(x-\frac{L}{2}\right)^2}}\right]\right]$	Potential respect to remote earth in any point in the ground due to an electrode infinitely deep. Two coordinate system.	1.4 (Figure 1-2)
$R = \frac{\rho}{2\pi}\cdot\frac{1}{L}\cdot\left[\ln\left[\frac{L}{a}\cdot\left[\sqrt{1+\left(\frac{a}{L}\right)^2}+1\right]\right]+\frac{a}{L}-\sqrt{1+\left(\frac{a}{L}\right)^2}\right]$	Resistance to remote earth of an electrode located infinitely deep in the ground.	1.8 (Figure 1-2)
$R = \frac{\rho}{2\pi\cdot L}\cdot\left(\ln\left(2\cdot\frac{L}{a}\right)-1\right)$	Resistance to remote earth of an electrode located infinitely deep in the ground when its diameter is <<< than its length.	1.9
$R = \frac{\rho}{2\cdot\pi}\cdot\frac{1}{L}\cdot\left[\ln\left[\frac{2\cdot L}{a}\cdot\left[\sqrt{1+\left(\frac{a}{2\cdot L}\right)^2}+1\right]\right]+\frac{a}{2\cdot L}-\sqrt{1+\left(\frac{a}{2\cdot L}\right)^2}\right]$	Resistance to remote earth of a vertical flush mounted electrode.	1.11 (Figure 1-4)
$R = \frac{\rho}{2\cdot\pi}\cdot\frac{1}{L}\cdot\left(\ln\left(\frac{4\cdot L}{a}\right)-1\right)$	Resistance to remote earth of a vertical flush mounted electrode when its diameter is <<< than its length.	1.12
$V(x,y) = \frac{I}{2\pi}\cdot\frac{\rho}{L}\cdot\ln\left(\frac{\sqrt{L^2+x^x+y^2}-L}{\sqrt{x^2+y^2}}\right)$	Potential respect to remote earth in any point in the ground surface due to an electrode flush mounted. Two coordinate system.	1.16 (Figure 1-5)
$V(x,y,z) = \frac{I}{2\pi}\cdot\frac{\rho}{L}\cdot\ln\left[\frac{L-z+\sqrt{x^2+y^2+(L-z)^2}}{-z+\sqrt{x^2+y^2+z^2}}\right]$	Potential respect to remote earth in any point in the ground due to an electrode flush mounted. Three coordinate system.	1.19 (Figure 1-7)
$V(r,z) = \frac{\rho}{4\cdot\pi}\cdot\frac{I}{L}\cdot\left[\ln\left[\frac{L+t+z+\sqrt{r^2+(z+L+t)^2}}{z+t+\sqrt{r^2+(z+t)^2}}\right]+\ln\left[\frac{z-t+\sqrt{r^2+(z-t)^2}}{z-t-L+\sqrt{r^2+(z-t-L)^2}}\right]\right]$	Potential respect to remote earth in any point in the ground due to an electrode at depth t from earth surface.	2.3 (Figure 2-1)

Equation	What is calculated?	Equation number
$V(r) = \frac{\rho}{4\cdot\pi}\cdot\frac{I}{L}\cdot\left[\ln\left[\frac{L + t + \sqrt{r^2 + (L+t)^2}}{t + \sqrt{r^2 + (t)^2}}\right] + \ln\left[\frac{-t + \sqrt{r^2 + t^2}}{-t - L + \sqrt{r^2 + (t+L)^2}}\right]\right]$	Potential respect to remote earth in any point in the ground surface due to a vertical electrode at depth t from earth surface.	2.4 (Figure 2-1)
$Rr1 := \frac{\rho}{4\pi}\cdot\frac{1}{L}\cdot\left[\ln\left(\frac{\sqrt{L^2 + a^2} + L}{\sqrt{L^2 + a^2} - L}\right) + 2\frac{(L+t)}{L}\cdot\ln\left[\frac{\sqrt{(L+2t)^2 + a^2} - (L+2t)}{\sqrt{4\cdot(L+t)^2 + a^2} - 2\cdot(L+t)}\right] + \frac{2\cdot t}{L}\cdot\ln\left[\frac{\sqrt{(L+2t)^2 + a^2} - (L+2t)}{\sqrt{(2\cdot t)^2 + a^2} - 2t}\right]\right]$	Resistance to remote earth of a vertical electrode at depth t from earth surface.	2.7
$Rr2 := \frac{\rho}{4\pi}\cdot\frac{1}{L}\cdot\left[\frac{2\sqrt{(L+2t)^2 + a^2}}{L} - \frac{\sqrt{2t^2 + a^2}}{L} - \frac{\sqrt{4(L+t)^2 + a^2}}{L} - \frac{2\sqrt{L^2 + a^2}}{L} + \frac{2a}{L}\right]$		
$Rr := Rr1 + Rr2$		
$V(x,y,z) = \frac{I}{4\pi}\cdot\frac{\rho}{L}\cdot\left[\ln\left[\frac{y + \sqrt{x^2 + y^2 + (z+t)^2}}{y - L + \sqrt{x^2 + (y-L)^2 + (z+t)^2}}\right] + \ln\left[\frac{y + \sqrt{x^2 + y^2 + (z-t)^2}}{y - L + \sqrt{x^2 + (y-L)^2 + (z-t)^2}}\right]\right]$	Potential respect to remote earth in any point in the ground surface due to a horizontal electrode at depth t from earth surface.	2.11.1 (Figure 2-2)
$Ug(x,y) := \frac{Ides}{2\pi}\cdot\frac{\rho}{L}\cdot\int_0^L \left[\frac{1}{\sqrt{(x - x1 - u\cdot\cos)^2 + (y - y1 - u\cdot\sin)^2 + t^2}}\right] du$	Potential in the ground respect to remote earth due to a horizontal electrode in an angle respect to the pipeline	2.15 (Figure 2-3)
$a(x,y) = \frac{\sqrt{L^2 - 2\cdot L\cdot[(x - x1)\cdot\cos + (y - y1)\cdot\sin] + (x - x1)^2 + (y - y1)^2 + t^2}}{\left[\sqrt{(x - x1)^2 + (y - y1)^2 + t^2} - [(x - x1)\cdot\cos + (y - y1)\cdot\sin]\right]}$	Potential in the ground respect to remote earth due to a horizontal electrode in an angle respect to the pipeline	2.16 (Figure 2-3)
$b(x,y) = \frac{L - [(x - x1)\cdot\cos + (y - y1)\cdot\sin]}{\left[\sqrt{(x - x1)^2 + (y - y1)^2 + t^2} - [(x - x1)\cdot\cos + (y - y1)\cdot\sin]\right]}$		
$Uae1(x,y) = \frac{Ides}{2\pi}\cdot\frac{\rho}{L}\cdot\ln(a(x,y) + b(x,y))$		
$V(x,y,z) = o\cdot\left[\sum_{n=0}^{N-1}\left[\ln\left[\frac{z - t + \sqrt{x^2 + (y - \xi - n\cdot S)^2 + (z-t)^2}}{z - t - L + \sqrt{x^2 + (y - \xi - n\cdot S)^2 + (z - t - L)^2}}\right] + \ln\left[\frac{t + L + z + \sqrt{x^2 + (y - \xi - n\cdot S)^2 + (t + L + z)^2}}{t + z + \sqrt{x^2 + (y - \xi - n\cdot S)^2 + (t+z)^2}}\right]\right]\right]$	Potential in the ground respect to remote earth due to a vertical groundbed, installed perpendicular to the pipeline.	3.4 (Figure 3-7 & 3-8)

Equation	What is calculated?	Equation number
$V(x,y,z) = \sigma \cdot \left[\sum_{n=0}^{N-1} \left[\ln\left[\frac{z - t + \sqrt{(x - n \cdot S)^2 + (y - \xi)^2 + (z - t)^2}}{z - t - L + \sqrt{(x - n \cdot S)^2 + (y - \xi)^2 + (z - t - L)^2}} \right] + \ln\left[\frac{t + L + z + \sqrt{(x - n \cdot S)^2 + (y - \xi)^2 + (t + L + z)^2}}{t + z + \sqrt{(x - n \cdot S)^2 + (y - \xi)^2 + (t + z)^2}} \right] \right] \right]$	Potential in the ground respect to remote earth due to a vertical groundbed, installed parallel to the pipeline.	3.5 (Figure 3-10 & 3-11)
$Vh(x,y,z) = \sigma \cdot \sum_{n=0}^{N-1} \left[\ln\left[\frac{x + \sqrt{x^2 + (y - \xi - n \cdot S)^2 + (t + z)^2}}{x - L + \sqrt{(x - L)^2 + (y - \xi - n \cdot S)^2 + (z + t)^2}} \right] + \ln\left[\frac{x + \sqrt{x^2 + (y - \xi - n \cdot S)^2 + (z - t)^2}}{x - L + \sqrt{(x - L)^2 + (y - \xi - n \cdot S)^2 + (z - t)^2}} \right] \right]$	Potential in the ground respect to remote earth due to a horizontal groundbed, installed perpendicular to the pipeline but each electrode is parallel to the pipeline.	3.7 (Figure 3-14)
$Vh(x,y,z) = \sigma \cdot \left[\sum_{n=0}^{N-1} \left[\ln\left[\frac{y - \xi - n \cdot S + \sqrt{x^2 + (y - \xi - n \cdot S)^2 + (t + z)^2}}{y - \xi - L - n \cdot S + \sqrt{x^2 + (y - \xi - L - n \cdot S)^2 + (t + z)^2}} \right] + \ln\left[\frac{y - \xi - n \cdot S + \sqrt{x^2 + (y - \xi - n \cdot S)^2 + (z - t)^2}}{y - \xi - L - n \cdot S + \sqrt{x^2 + (y - \xi - L - n \cdot S)^2 + (z - t)^2}} \right] \right] \right]$	Potential in the ground respect to remote earth due to a horizontal ground-bed, installed perpendicular to the pipeline and each electrode is perpendicular as well.	3.8 (Figure 3-15)
$Vh(x,y,z) = \sigma \cdot \left[\sum_{n=0}^{N-1} \left[\ln\left[\frac{y - \xi + \sqrt{(x - n \cdot S)^2 + (y - \xi)^2 + (t + z)^2}}{y - \xi - L + \sqrt{(x - n \cdot S)^2 + (y - \xi - L)^2 + (t + z)^2}} \right] + \ln\left[\frac{y - \xi + \sqrt{(x - n \cdot S)^2 + (y - \xi)^2 + (z - t)^2}}{y - L - \xi + \sqrt{(x - n \cdot S)^2 + (y - \xi - L)^2 + (z - t)^2}} \right] \right] \right]$	Potential in the ground respect to remote earth due to a horizontal ground-bed, installed parallel to the pipeline and each electrode is perpendicular to the pipeline.	3.10 (Figure 3-16)
$Vh(x,y,z) = \sum_{n=0}^{N-1} \left[\sigma \cdot \left[\ln\left[\frac{x - n \cdot S + \sqrt{(x - n \cdot S)^2 + (y - \xi)^2 + (z - t)^2}}{x - L - n \cdot S + \sqrt{(x - L - n \cdot S)^2 + (y - \xi)^2 + (z - t)^2}} \right] + \ln\left[\frac{x - n \cdot S + \sqrt{(x - n \cdot S)^2 + (y - \xi)^2 + (z + t)^2}}{x - L - n \cdot S + \sqrt{(x - L - n \cdot S)^2 + (y - \xi)^2 + (z + t)^2}} \right] \right] \right]$	Potential in the ground respect to remote earth due to a horizontal ground-bed, installed parallel to the pipeline and each electrode is parallel as well	3.12 (Figure 3-17)

Equation	What is calculated?	Equation number
$Ui1(x,y,z)=\sum_{n=0}^{N-1}\left[o\cdot\ln\left[\frac{t+L+z+\sqrt{(x-n\cdot S\cdot coseno-x1)^2+(y-n\cdot S\cdot seno-y1)^2+(t+L+z)^2}}{z+t+\sqrt{(x-n\cdot S\cdot coseno-x1)^2+(y-n\cdot S\cdot seno-y1)^2+(z+t)^2}}\right]\right]$ $Ui2(x,y,z)=\sum_{n=0}^{N-1}\left[o\cdot\ln\left[\frac{z-t+\sqrt{(x-n\cdot S\cdot coseno-x1)^2+(y-n\cdot S\cdot seno-y1)^2+(z-t)^2}}{z-t-L+\sqrt{(x-n\cdot S\cdot coseno-x1)^2+(y-n\cdot S\cdot seno-y1)^2+(z-t-L)^2}}\right]\right]$ $Ui(x,y,z)=Ui1(x,y,z)+Ui2(x,y,z)$	Potential in the ground due to a vertical groundbed. The groundbed is at any angle respect to the pipeline.	3.14
$Uh1(x,y,z)=\sum_{p=0}^{N-1}\left[o\cdot\ln\left[\frac{\sqrt{(x-p\cdot S\cdot coseno-x1)^2+(y-p\cdot S\cdot seno-y1)^2+(z-t)^2}+(x-p\cdot S\cdot coseno-x1)}{\sqrt{(x-p\cdot S\cdot coseno-x1-L)^2+(y-p\cdot S\cdot seno-y1)^2+(z-t)^2}+(x-L-p\cdot S\cdot coseno-x1)}\right]\right]$ $Uh2(x,y,z)=\sum_{p=0}^{N-1}\left[o\cdot\ln\left[\frac{\sqrt{(x-p\cdot S\cdot coseno-x1)^2+(y-p\cdot S\cdot seno-y1)^2+(z+t)^2}+(x-p\cdot S\cdot coseno-x1)}{\sqrt{(x-p\cdot S\cdot coseno-x1-L)^2+(y-p\cdot S\cdot seno-y1)^2+(z+t)^2}+(x-L-p\cdot S\cdot coseno-x1)}\right]\right]$ $Uh(x,y,z)=Uh1(x,y,z)+Uh2(x,y,z)$	Horizontal groundbed, where the electrodes are in discrete coke beds, parallel to the pipeline and groundbed is at an angle from the pipeline.	3.15 (Figure 3-18)
$Uh1(x,y,z)=\sum_{p=0}^{N-1}\left[o\cdot\ln\left[\frac{\sqrt{(x-p\cdot S\cdot coseno-x1)^2+(y-p\cdot S\cdot seno-y1)^2+(z-t)^2}+(y-p\cdot S\cdot seno-y1)}{\sqrt{(x-p\cdot S\cdot coseno-x1)^2+(y-L-p\cdot S\cdot seno-y1)^2+(z-t)^2}+(y-L-p\cdot S\cdot seno-y1)}\right]\right]$ $Uh2(x,y,z)=\sum_{p=0}^{N-1}\left[o\cdot\ln\left[\frac{\sqrt{(x-p\cdot S\cdot coseno-x1)^2+(y-p\cdot S\cdot seno-y1)^2+(z+t)^2}+(y-p\cdot S\cdot seno-y1)}{\sqrt{(x-p\cdot S\cdot coseno-x1)^2+(y-L-p\cdot S\cdot seno-y1)^2+(z+t)^2}+(y-L-p\cdot S\cdot seno-y1)}\right]\right]$ $Uh(x,y,z)=Uh1(x,y,z)+Uh2(x,y,z)$	Horizontal groundbed, where the electrodes are in discrete coke beds, perpendicular to the pipeline and groundbed is at any angle from the pipeline.	3.16 (Figure 3-19)
$Ug1(x,y,z)=o\cdot\left[\sum_{p=0}^{N-1}\int_0^L\left[\frac{1}{\sqrt{[x-x1-(u+p\cdot Ln)\cdot coseno]^2+[y-y1-(u+p\cdot Ln)\cdot seno]^2+(z-t)^2}}\right]du\right]$ $Ug2(x,y,z)=o\cdot\left[\sum_{p=0}^{N-1}\int_0^L\left[\frac{1}{\sqrt{[x-x1-(u+p\cdot Ln)\cdot coseno]^2+[y-y1-(u+p\cdot Ln)\cdot seno]^2+(z+t)^2}}\right]du\right]$ $Ug(x,y,z)=Ug1(x,y,z)+Ug2(x,y,z)$	Horizontal groundbed, where the electrodes are in discrete coke beds, in line and groundbed is at any angle from the pipeline.	3.17 (Figure 3-20)
$UaeS(x,y,z)=o\cdot\sum_{p=0}^{N-1}\left(\ln\left(\frac{a1(x,y,z,p)+b1(x,y,p)}{c1(x,y,z,p)}\right)+\ln\left(\frac{a2(x,y,z,p)+b1(x,y,p)}{c2(x,y,z,p)}\right)\right)$	Horizontal groundbed, where the electrodes are in discrete coke beds, in line, and groundbed is at an angle from the pipeline. Analytical solution of Equation 17.	3.19 (Figure 3-20)
$Rg=\frac{1}{N}\cdot\left[\frac{1}{2\cdot\pi}\cdot\frac{\rho}{L}\cdot\left(\ln\left(4\cdot\frac{L}{a}\right)-1\right)+\frac{2}{N}\cdot\left[\sum_{n=1}^{N-1}[(N-n)\cdot R(nS)]\right]\right]$	Resistance of vertical groundbed, general form.	4.9

Equation	What is calculated?	Equation number
$R0=\frac{\rho}{2\pi\cdot L}\cdot\left[\ln\left(4\cdot\frac{L}{d}\right)-1+\ln\left[\frac{L+\sqrt{(2\cdot t)^2+L^2}}{2\cdot t}\right]+\frac{2\cdot t}{L}-\frac{\sqrt{(2\cdot t)^2+(L)^2}}{L}\right]$	Resistance equation for horizontal groundbed with electrodes in discrete coke beds.	2.14 (Figure 4-1, 4-2 & 4-3)
$R1=\frac{\rho}{2\pi}\cdot\frac{1}{L}\cdot\frac{2}{N}\cdot\left[\sum_{m=1}^{N-1}\left[(N-m)\cdot\left[\ln\left[\frac{L+\sqrt{(mS)^2+L^2}}{mS}\right]+\frac{mS}{L}-\frac{\sqrt{(mS)^2+(L)^2}}{L}\right]\right]\right]$		4.13
$R2=\frac{\rho}{2\pi}\cdot\frac{1}{L}\cdot\frac{2}{N}\cdot\sum_{m=1}^{N-1}\left[(N-m)\cdot\left[\ln\left[\frac{L+\sqrt{(mS)^2+(2t)^2+L^2}}{\sqrt{(mS)^2+(2t)^2}}\right]+\frac{\sqrt{(mS)^2+(2t)^2}}{L}-\frac{\sqrt{(mS)^2+(2t)^2+(L)^2}}{L}\right]\right]$		4.14
$Rh=\frac{R0+R1+R2}{N}$		4.15
$RhL=\frac{R0}{N}$	Resistance equation for horizontal groundbe with electrodes in line in discrete coke beds.	4.16 (Figure 4-4)
$Resist=\frac{1}{\sum_{n=1}^{N}\left[\frac{1}{\left[\left[\frac{1}{N}\cdot\sum_{i=1}^{N}(Ga_i)+\sum_{i=1}^{N-1}(Gi_i)\right]\right]}\right]}$	Resistance equation for vertical electrodes with dissimilar length and depth below grade.	5.7 (Figure 5-1)
$Rgb=\frac{1}{N}\cdot\left[\frac{1}{N}\cdot\left[\sum_{i=1}^{N}(Ga_i)\right]+\left[\sum_{i=1}^{N-1}(Gi_i)\right]\right]$	Resistance equation for vertical electrodes with dissimilar length and depth below grade.	5.8 (Figure 5-1)
$U(x,y,z)=\sum_{j=1}^{N}\left[\left(\frac{I\cdot\rho}{4\cdot\pi\cdot N\cdot L_j}\right)\left[\int_{-L_j-t_j}^{-t_j}\left[\frac{1}{\sqrt{x^2+[y-(j-1)\cdot S]^2+(z-u)^2}}\right]du+\int_{t_j}^{L_j+t_j}\left[\frac{1}{\sqrt{x^2+[y-(j-1)\cdot S]^2+(z-u)^2}}\right]du\right]\right]$	Potential to remote earth due to groundbed with vertical electrodes with dissimilar length and depth below grade, intergral form.	5.16 (Figure 5-1)
$U(x,y,z)=\sum_{j=0}^{N-1}\left[o_j\cdot\left[\ln\left[\frac{z-t_j+\sqrt{x^2+(y-j\cdot S)^2+(z-t_j)^2}}{z-t_j-L_j+\sqrt{x^2+(y-j\cdot S)^2+(z-t_j-L_j)^2}}\right]+\ln\left[\frac{t_j+L_j+z+\sqrt{x^2+(y-j\cdot S)^2+(t_j+L_j+z)^2}}{t_j+z+\sqrt{x^2+(y-j\cdot S)^2+(t_j+z)^2}}\right]\right]\right]$	Potential to remote earth due to groundbed with vertical electrodes with dissimilar length and depth below grade, analytical form.	5.17 (Figure 5-1)

Equation	What is calculated?	Equation number
$Rgb = \frac{1}{N}\cdot\left[\frac{1}{N}\cdot\left[\sum_{i=1}^{N}(Ga_i)\right]+\left[\sum_{i=1}^{N}(Gi_i)\right]\right]$	Resistance equation for vertical electrodes with same length and depth and dissimilar distance between electrodes.	5.25 (Figure 5-2)
$Resist = \frac{1}{\sum_{n=1}^{N}\left[\frac{1}{\left[\left[\frac{1}{N}\cdot\sum_{i=1}^{N}(Ga_i)+\sum_{i=1}^{N}(Gi_i)\right]\right]}\right]}$	Resistance equation for vertical electrodes with same length and depth and dissimilar distance between electrodes.	5.26 (Figure 5-2)
$Vds1(x,y,z) = \frac{1}{N}\cdot\left(\frac{\rho}{4\cdot\pi}\cdot\frac{1}{L}\right)\left[\ln\left[\frac{L+t+z+\sqrt{x^2+(y-\xi)^2+(z+L+t)^2}}{z+t+\sqrt{x^2+(y-\xi)^2+(z+t)^2}}\right]+\ln\left[\frac{z-t+\sqrt{x^2+(y-\xi)^2+(z-t)^2}}{z-t-L+\sqrt{x^2+(y-\xi)^2+(z-t-L)^2}}\right]\right]$	Potential equation for vertical electrodes with same length and depth and dissimilar distance between electrodes.	5.27 (Figure 5-2 and 5-3)
$Vds2(x,y,z) = \frac{1}{N}\cdot\left(\frac{\rho}{4\cdot\pi}\cdot\frac{1}{L}\right)\sum_{j=1}^{N-1}\left[\ln\left[\frac{L+t+z+\sqrt{x^2+\left(y-\sum_{n=1}^{j}S_n-\xi\right)^2+(z+L+t)^2}}{z+t+\sqrt{x^2+\left(y-\sum_{n=1}^{j}S_n-\xi\right)^2+(z+t)^2}}\right]+\ln\left[\frac{z-t+\sqrt{x^2+\left(y-\sum_{n=1}^{j}S_n-\xi\right)^2+(z-t)^2}}{z-t-L+\sqrt{x^2+\left(y-\sum_{n=1}^{j}S_n-\xi\right)^2+(z-t-L)^2}}\right]\right]$		5.28
$Vdm(x,y,z) = Vds1(x,y,z) + Vds2(x,y,z)$		5.29
$Rs = \frac{1}{2N}\cdot\left[RL + RLS + \frac{2}{N}\cdot\left[\sum_{i=1}^{N-1}\left[(N-i)\cdot RLt_i\right]\right]+\frac{2}{N}\cdot\left[\sum_{i=1}^{N-1}\left[(N-i)\cdot RLtt_i\right]\right]\right]$	Resistance equation for two vertical groundbeds in parallalel.	5.31, equation components in the text (Figure 5-4)
$U1(x,y,z) = \sum_{n=0}^{N-1}\left[\sigma\cdot\left[\ln\left[\frac{z+t+L+\sqrt{(x-\beta)^2+(y-n\cdot S-\xi)^2+(z+t+L)^2}}{z+t+\sqrt{(x-\beta)^2+(y-n\cdot S-\xi)^2+(z+t)^2}}\right]+\ln\left[\frac{z-t+\sqrt{(x-\beta)^2+(y-n\cdot S-\xi)^2+(z-t)^2}}{z-t-L+\sqrt{(x-\beta)^2+(y-n\cdot S-\xi)^2+(z-t-L)^2}}\right]\right]\right]$	Potential equation for two vertical groundbeds in parallel.	5.32 (Figure 5-4, 5-5)
$U2(x,y,z) = \sum_{n=0}^{N-1}\left[\sigma\cdot\left[\ln\left[\frac{z+t+L+\sqrt{(x-S1-\beta)^2+(y-n\cdot S-\xi)^2+(z+t+L)^2}}{z+t+\sqrt{(x-S1-\beta)^2+(y-n\cdot S-\xi)^2+(z+t)^2}}\right]+\ln\left[\frac{z-t+\sqrt{(x-S1-\beta)^2+(y-n\cdot S-\xi)^2+(z-t)^2}}{z-t-L+\sqrt{(x-S1-\beta)^2+(y-n\cdot S-\xi)^2+(z-t-L)^2}}\right]\right]\right]$		5.33
$U(x,y,z) = U1(x,y,z) + U2(x,y,z)$		5.34

Equation	What is calculated?	Equation number
$UT1(x,y,z) = \sigma 1 \cdot \left[\ln \left[\frac{x - Xw + \sqrt{(x - Xw)^2 + (y - Yw)^2 + (t + z)^2}}{x - Xw - L + \sqrt{(x - L - Xw)^2 + (y - Yw)^2 + (z + t)^2}} \right] \right]$	Potential equation when the electrodes are distributed non-uniformly.	5.37, variables explanation in the text (Figure 5-6)
$UT2(x,y,z) = \sigma 1 \cdot \left[\ln \left[\frac{x - Xw + \sqrt{(x - Xw)^2 + (y - Yw)^2 + (z - t)^2}}{x - L - Xw + \sqrt{(x - L - Xw)^2 + (y - Yw)^2 + (z - t)^2}} \right] \right]$		
$UT(x,y,z) = UT1(x,y,z) + UT2(x,y,z)$		
$Uts(x,y,z) = \sum_{n=1}^{rows(Xw)} UT(x,y,z)_n$		
$En1_{y,n} = -\left[E0_n \cdot \cosh\left[\alpha p_n \cdot (y - \xi n1) \right] - Rgp_n \cdot In1_n \cdot \sinh\left[-\alpha p_n \cdot (y - \xi n1) \right] \right]$	Attenuation equation of a pipeline including the negative connection for the case of multiple rectifier systems.	6.5, variables explanation in the text (Figure 6-6, 6-7)
$EnOn1_{y,n} = EN_{y,n} + Ens1_{y,n} + Uperp1_{y,n}$	Transmission line "On" potential equation.	6.7, variables explanation in the text (Figure 6-6, 6-7)
$I = \frac{2 \cdot \pi \cdot \Delta Ux}{\rho 0} \cdot \frac{t \cdot \sqrt{t^2 + x^2}}{\sqrt{t^2 + x^2} - t}$	Current at an anomaly.	6.9
$Ro = \frac{\Delta Ur}{I}$	Resistance of the anomaly.	6.10
$ro = \frac{\pi}{2} \cdot \frac{\Delta Ux}{\Delta Ur} \cdot \frac{t \cdot \sqrt{t^2 + x^2}}{\sqrt{t^2 + x^2} - t}$	Anomaly radius.	6.11
$\delta I = \frac{I}{\pi \cdot ro^2}$	Current density at the anomaly.	6.12

Introduction to the Use of the Average Potential Method (APM) for the Calculation of Resistance for an Electrode in Soil

In this chapter, the methodology to obtain the classical equations for electrode potential and resistances of underground electrodes, i.e., grounding and CP groundbed systems is discussed and calculations developed to facilitate the reader's understanding. This chapter is the foundation for the following chapters dedicated to potential and resistance calculation of buried electrodes in different configurations.

Key features within this chapter are:

- The understanding of the coordinate system and its origin and use to obtain the formulations, presented.
- Calculate the distance between the differential charge element in the electrode axis and the point in the space where the potential is determined.

These two steps are critical to the interpretation of the data obtained when the calculated values are compared with the actual potential values measured in the field, in practical applications.

1.1 Potential calculation of an electrode in the space of infinite extent.

The development below determines the potential to remote earth, generated by the current $2I(0)$, entering the electrode at its center $(x=0, y=0, z=0)$. The electrode is in space of infinite extent, where no borders or charged bodies are considered affecting the shape of the equipotential lines generated by the current leaving the electrode surface.[1]

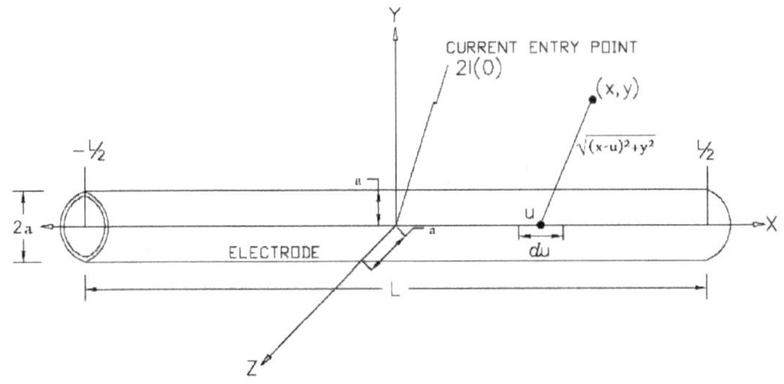

Figure 1-1: Potential calculation due to an electrode in infinite space

This electrode could be an impressed current anode located deep in the ground.

Our objective here is to develop the integration steps that may have been incorporated by Sunde to obtain his equations. We will also attempt to explain the assumptions that are behind these equations.

Part of the physical and mathematical assumptions required are:

- The current distributes uniformly in the electrode length.
- The current is entering the electrode at its mid-point.
- The current concentrates in the electrode's axis.
- The electrode length L is much larger than its radius a.
- The origin of coordinates for integration is at the electrode mid-point.
- The integration variable is u that runs in the axis of the electrode, from $-L/2$ to $L/2$.
- The underground electrode is in direct contact with its surroundings. This assumes that there are no air gaps between the surface of the buried electrode and the soil[2]. This is not completely true, even after grounds electrodes have been installed for a long time, and that, as a result of contact resistance at the surface of the conductor, the resistance of grounds may be some 20% higher than calculated.

The concept of the charge concentrated in the electrode's axis is derived from the Newtonian potential theory[3].

The mathematical explanation of the calculations is as follows:

The axis of the electrode is on the x-axis. The variable u is used to perform the integration; this variable is running between the ends of the electrode: $-L/2$ and $L/2$. This mathematical development allows one the advantage to keep the potential as a function of the variables x and y for future calculations.

First determine the potential at point *(x,y,z)* in the space due to a conductor element *du* centered at the point *x=u, y=0* and *z=0* in the axis of the electrode, with current *dI(u)*

Due to the electrode is remote from the surface of the earth, the point on the electrode at coordinates *(u,0,0)* radiates current in all directions in a spherical-shaped equipotential surfaces[4]. The current density due to its current *dI(u)*, at the point *(x,y,z)* is determined dividing the current by the surface area of a sphere with radius equal to the distance between the point charge at *(u,0,0)* and the point *(x,y)*. Figure 1-2.

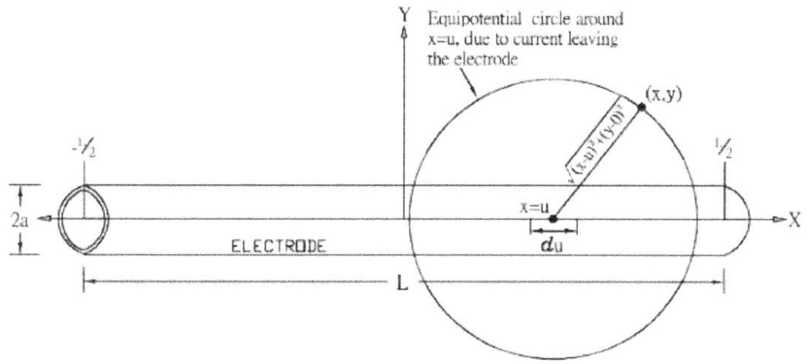

Figure 1-2: Potential calculation at point (x,y)

The distance between the electrode element *du* at *(u,0,0)* and *(x,y,z)* is:

$$r = \sqrt{(x-u)^2 + (y-0)^2 + (z-0)^2} \qquad \text{Equation 1.1}$$

Because the calculations are only related to the *x*-axis, and to simplify, the *y* and *z* axes are condensed in the *y*-axis only, so:

$$y = \sqrt{y^2 + z^2}$$

The variable *y* will represent afterward the radius of the electrode *a*, and will correspond to the value for the *y* and *z* spatial axes, as can be seen in the Figure 1-2

The sphere surface area with radius r is:

$$4 \cdot \pi \cdot r^2$$

The current density at *(x,y)* due to current *dI(u)* in the electrode element *du* is then:

$$\delta I(x,y) = dI(u)\cdot\frac{1}{\left(4\cdot\pi\cdot r^2\right)}$$

If ρ is the soil resistivity around the electrode, the electric intensity at point *(x,y)* will be:

$$E(x,y) = dI(u)\cdot\frac{\rho}{4\cdot\pi\cdot r^2}$$

This is the intensity of the electric field developed by the differential charge *du* at the point *(x,y)*.

And the potential at the point *(x,y)* due to the current at the element du is:

$$\frac{dV(r)}{du} = -\int_{\infty}^{r}\frac{\rho}{4\cdot\pi\cdot r^2}\cdot\frac{dI(u)}{du}\,dr$$

Calculating the integral, we obtain:

$$\frac{dV(r)}{du} = \frac{dI(u)}{du}\cdot\frac{\rho}{4\cdot\pi\cdot r}$$

Now substituting the *r* from Equation 1.1 above, and leaving the potential differential unchanged, we obtain:

$$dV(x,y) = \frac{dI(u)}{du}\cdot\frac{\rho}{4\cdot\pi\cdot\sqrt{(x-u)^2+(y)^2}}\cdot du \qquad \text{Equation 1.2}$$

If the current distributes uniformly along the axis of the electrode, we can substitute the following:

$$\frac{dI(u)}{du} = \frac{2\cdot I(0)}{L}$$

Where *2I(0)/L* is the average leakage current density when current *2I(0)* is injected into the center of the electrode and *L* is the length of the electrode. We can also consider that the current does not attenuate as a function of the electrode's length.

Hence, the potential due to *du*, over the point *(x,y)*, can be integrated using Equation 1.2, over the entire length of the electrode. This will provide the potential due to the influence of the entire electrode at the point *(x,y)*. This can be done considering the additive effect, which means that the potential at a point due to a group of charges will be the sum of the potential due to each charge. This is performed using a linear integral over the entire length of the charged electrode.

For this integration, the origin of coordinates is located at the center of the electrode.

The integral will have the form:

$$V(x,y) = \frac{\rho}{4\cdot\pi}\cdot\frac{2\cdot I(0)}{L}\cdot\int_{\frac{-L}{2}}^{\frac{L}{2}} \frac{1}{\sqrt{(x-u)^2+y^2}}\,du \qquad \text{Equation 1.3}$$

Now this integral is calculated as follow:

$$V(x,y) = \frac{\rho}{4\cdot\pi}\cdot\frac{2\cdot I(0)}{L}\cdot\left[\int_{\frac{-L}{2}}^{\frac{L}{2}} \frac{1}{y\cdot\sqrt{\left(\frac{x-u}{y}\right)^2+1}}\,du\right]$$

Substituting these variables:

$$\frac{x-u}{y} = \theta \qquad d\theta = \frac{-du}{y}$$

The new limits for integration are:

$$u = \frac{-L}{2} \qquad \theta = \frac{x+\frac{L}{2}}{y}$$

$$u = \frac{L}{2} \qquad \theta = \frac{x-\frac{L}{2}}{y}$$

And the integral is:

$$V(x,y) = \frac{\rho}{4\cdot\pi}\cdot\frac{2\cdot I(0)}{L}\cdot\left[\int_{\frac{x-\frac{L}{2}}{y}}^{\frac{x+\frac{L}{2}}{y}} \frac{1}{\sqrt{(\theta)^2+1}}\,d\theta\right]$$

The general solution for this integral is as follow (without integration limits)

$$V(x,y) = \frac{\rho}{4\cdot\pi}\cdot\frac{2\cdot I(0)}{L}\cdot\ln\left(\theta+\sqrt{1+\theta^2}\right)$$

When the integration limits are considered the result is:

$$V(x,y) = \frac{\rho}{4\cdot\pi}\cdot\frac{2\cdot I(0)}{L}\cdot\left[\ln\left[\frac{x+\frac{L}{2}}{y}+\sqrt{1+\left(\frac{x+\frac{L}{2}}{y}\right)^2}\right]-\ln\left[\frac{x-\frac{L}{2}}{y}+\sqrt{1+\left(\frac{x-\frac{L}{2}}{y}\right)^2}\right]\right]$$

Now using properties of logarithms:

$$V(x,y) = \frac{\rho}{4\cdot\pi}\cdot\frac{2\cdot I(0)}{L}\cdot\left[\ln\left[\frac{\frac{x+\frac{L}{2}}{y}+\sqrt{1+\left(\frac{x+\frac{L}{2}}{y}\right)^2}}{\frac{x-\frac{L}{2}}{y}+\sqrt{1+\left(\frac{x-\frac{L}{2}}{y}\right)^2}}\right]\right]$$

Taking *1/y* as a common factor the new form is:

$$V(x,y) = \frac{\rho}{4\cdot\pi}\cdot\frac{2\cdot I(0)}{L}\cdot\left[\ln\left[\frac{x+\frac{L}{2}+\sqrt{y^2+\left(x+\frac{L}{2}\right)^2}}{x-\frac{L}{2}+\sqrt{y^2+\left(x-\frac{L}{2}\right)^2}}\right]\right]$$ Equation 1.4

Now, the potential at the surface of the electrode *(y=a)* is:

$$V(x,a) = \frac{\rho}{4\cdot\pi}\cdot\frac{2\cdot I(0)}{L}\cdot\left[\ln\left[\frac{x+\frac{L}{2}+\sqrt{a^2+\left(x+\frac{L}{2}\right)^2}}{x-\frac{L}{2}+\sqrt{a^2+\left(x-\frac{L}{2}\right)^2}}\right]\right]$$ Equation 1.5

This is the potential of the electrode surface respect to remote earth, due to the charges on the electrode axis. Now, we are ready to obtain the electrode resistance to remote earth.

1.2 Resistance to remote earth for an electrode in the space of infinite extent

The evaluation of the potential at the electrode surface (Equation 1.5) above is critical for us to calculate the resistance of the electrode to remote earth.

Until now, we have the potential due to current *2I(0)* emanating from the electrode axis over a point in the electrode surface: *(x,a)*, where ***a*** is the radius of the electrode.

The potential over the entire surface of the electrode due to the current can be obtained using the Average Potential Method (APM). The APM considers the total potential over the electrode surface to be the sum of the potential calculated per point (*(x,a)* in our case) divided by the length of the electrode. The potential of the entire electrode is considered uniform. If you sum all the potentials due to individual points and divide by the total points (i.e. the length of the electrode) you will have the average potential at the surface of the electrode.

The resistance of the electrode to remote earth is calculated dividing the average potential by the total current.

If we sum the potential obtained due to the total current at a point (Equation 1.5), integrate that Equation from *x=0* to *x=L/2*, and divide by the length of that section *L/2*, we will calculate the average potential in the entire electrode. This is the same if we calculate the potential integrating from *x=-L/2* and *x=L/2* and divide by *L*. In either case, we will obtain the same value, *the average potential.*

Then, the integration process is as follow:

$$V = \int_0^{\frac{L}{2}} \frac{\rho}{4\cdot\pi}\cdot\frac{2\cdot I(0)}{L}\cdot \ln\left[\left[\frac{x+\frac{L}{2}+\sqrt{a^2+\left(x+\frac{L}{2}\right)^2}}{x-\frac{L}{2}+\sqrt{a^2+\left(x-\frac{L}{2}\right)^2}}\right]\right] dx$$

By logarithm properties, this integral can be arranged as:

$$V=\frac{\rho}{4\cdot\pi}\cdot\frac{2\cdot I(0)}{L}\cdot\left(\int_0^{\frac{L}{2}} \ln\left(x+\frac{L}{2}+\sqrt{a^2+\left(x+\frac{L}{2}\right)^2}\right)dx-\int_0^{\frac{L}{2}} \ln\left(x-\frac{L}{2}+\sqrt{a^2+\left(x-\frac{L}{2}\right)^2}\right)dx\right)$$

We then separate the integral in two components to facilitate the calculation:

$$V=\frac{\rho}{4\cdot\pi}\cdot\frac{2\cdot I(0)}{L}\cdot\left(\int_0^{\frac{L}{2}} \ln\left(x+\frac{L}{2}+\sqrt{a^2+\left(x+\frac{L}{2}\right)^2}\right)dx-\int_0^{\frac{L}{2}} \ln\left(x-\frac{L}{2}+\sqrt{a^2+\left(x-\frac{L}{2}\right)^2}\right)dx\right)$$

Equation 1.6

$$A=\int_0^{\frac{L}{2}} \ln\left(x+\frac{L}{2}+\sqrt{a^2+\left(x+\frac{L}{2}\right)^2}\right)dx \quad B=\int_0^{\frac{L}{2}} \ln\left(x-\frac{L}{2}+\sqrt{a^2+\left(x-\frac{L}{2}\right)^2}\right)dx$$

$$V=\frac{\rho}{4\cdot\pi}\cdot\frac{2\cdot I(0)}{L}\cdot(A-B)$$

A and *B* are solved in the same manner. Equation 1.6 is applied to obtain the solution below.

By substitution of these variables:

$$\theta = x + \frac{L}{2} \qquad d\theta = dx$$

When $x = 0$ $\quad \theta = \frac{L}{2}$

When $x = \frac{L}{2}$ $\quad \theta = L$

We obtain:

$$A = \int_{\frac{L}{2}}^{L} \ln\left(\theta + \sqrt{a^2 + \theta^2}\right) d\theta$$

Substituting the variables below:

$$\theta = \sinh(\xi)\cdot a \qquad d\theta = a\cdot\cosh(\xi)\cdot d\xi$$

When, $\theta = \frac{L}{2}$ $\quad \xi = \operatorname{arcsh}\left(\frac{L}{2a}\right)$

When, $\theta = L$ $\quad \xi = \operatorname{arcsh}\left(\frac{L}{a}\right)$

By incorporating the hyperbolic functions properties, results in the following:

$$A = \int_{\operatorname{arcsh}\left(\frac{L}{2a}\right)}^{\operatorname{arcsh}\left(\frac{L}{a}\right)} \ln\left[\sinh(\xi)\cdot a + \sqrt{a^2 + (\sinh(\xi)\cdot a)^2}\right]\cdot a\cdot\cosh(\xi)\, d\xi$$

$$\cosh(\xi)^2 - \sinh(\xi)^2 = 1 \qquad \cosh(\xi)^2 = 1 + \sinh(\xi)^2$$

$$\frac{d}{d\xi}(\sinh(\xi)) = \cosh(\xi) \qquad \frac{d}{d\xi}(\cosh(\xi)) = \sinh(\xi)$$

$$A = \int_{\operatorname{arcsh}\left(\frac{L}{2a}\right)}^{\operatorname{arcsh}\left(\frac{L}{a}\right)} \ln(\sinh(\xi)\cdot a + a\cdot\cosh(\xi))\cdot a\cdot\cosh(\xi)\, d\xi$$

By substitution of variables:

$$dv = a\cdot\cosh(\xi)\cdot d\xi \quad \text{and} \quad u = \ln(\sinh(\xi)\cdot a + a\cdot\cosh(\xi))$$

And integrating by parts (Note: The integration limits have been kept apart until the integral is solved):

$$A = \ln[a\cdot(\sinh(\xi) + \cosh(\xi))]\cdot(\sinh(\xi)\cdot a) - \int \frac{\sinh(\xi)\cdot a\cdot a\cdot(\sinh(\xi) + \cosh(\xi))}{a\cdot(\sinh(\xi) + \cosh(\xi))}\, d\xi$$

$$A = \ln[a\cdot(\sinh(\xi) + \cosh(\xi))]\cdot(\sinh(\xi)\cdot a) - \int \sinh(\xi)\cdot a\, d\xi$$

$$A = \ln[a\cdot(\sinh(\xi) + \cosh(\xi))]\cdot(\sinh(\xi)\cdot a) - a\cdot\cosh(\xi)$$

$$A = \ln\left[a\cdot\left(\sinh(\xi) + \sqrt{1 + \sinh(\xi)^2}\right)\right]\cdot(\sinh(\xi)\cdot a) - a\cdot\left(\sqrt{1 + \sinh(\xi)^2}\right)$$

and by substituting $\theta = \sinh(\xi)\cdot a$ the following result is:

$$A = \left(\ln\left(\theta + \sqrt{a^2 + \theta^2}\right)\right)\cdot\theta - \sqrt{a^2 + \theta^2} \quad \text{and} \quad \theta = x + \frac{L}{2}$$

$$A = \left[\ln\left[x + \frac{L}{2} + \sqrt{a^2 + \left(x + \frac{L}{2}\right)^2}\right]\cdot\left(x + \frac{L}{2}\right) - \sqrt{a^2 + \left(x + \frac{L}{2}\right)^2}\right.$$

Now applying the integration limits:

$$x = 0 \quad \text{and} \quad x = \frac{L}{2}$$

$$A = \left[\ln\left[L + \sqrt{a^2 + (L)^2}\right]\cdot(L) - \sqrt{a^2 + (L)^2} - \left[\ln\left[\frac{L}{2} + \sqrt{a^2 + \left(\frac{L}{2}\right)^2}\right]\cdot\left(\frac{L}{2}\right) - \sqrt{a^2 + \left(\frac{L}{2}\right)^2}\right]\right]$$

Implementing the same procedure for *B*,

$$B = \left[\ln\left[x - \frac{L}{2} + \sqrt{a^2 + \left(x - \frac{L}{2}\right)^2}\right]\cdot\left(x - \frac{L}{2}\right) - \sqrt{a^2 + \left(x - \frac{L}{2}\right)^2}\right.$$

$$B = -a + \left[\ln\left[-\frac{L}{2} + \sqrt{a^2 + \left(-\frac{L}{2}\right)^2}\right]\cdot\left(-\frac{L}{2}\right) + \sqrt{a^2 + \left(-\frac{L}{2}\right)^2}\right.$$

Now, substituting *A* and *B* into the Equation 1.6:

$$V = \frac{\rho}{4\cdot\pi}\cdot\frac{2\cdot I(0)}{L}\cdot\left[L\cdot\left[\ln\left(\sqrt{L^2 + a^2} + L\right) - \ln(a) + \frac{a}{L} - \sqrt{1 + \left(\frac{a}{L}\right)^2}\right]\right]$$

By logarithm properties:

$$V = \frac{\rho}{4\cdot\pi}\cdot\frac{2\cdot I(0)}{L}\cdot L\cdot\left[\ln\left[\sqrt{\left(\frac{L}{a}\right)^2 + 1} + \frac{L}{a}\right] + \frac{a}{L} - \sqrt{1 + \left(\frac{a}{L}\right)^2}\right]$$

Now, dividing by $L/2$, to calculate the average potential.

Since the potential is uniform over the entire electrode, the average in one-half will be the same as for the entire electrode.

$$V = \frac{\rho}{2\cdot\pi}\cdot\frac{2\cdot I(0)}{L}\cdot\left[\ln\left[\sqrt{\left(\frac{L}{a}\right)^2 + 1} + \frac{L}{a}\right] + \frac{a}{L} - \sqrt{1+\left(\frac{a}{L}\right)^2}\right]$$

Having all the terms inside the Equation as a/L, we obtain:

$$V = \frac{\rho}{2\cdot\pi}\cdot\frac{2\cdot I(0)}{L}\cdot\left[\ln\left[\frac{L}{a}\cdot\sqrt{1+\left(\frac{a}{L}\right)^2} + 1\right] + \frac{a}{L} - \sqrt{1+\left(\frac{a}{L}\right)^2}\right] \qquad \text{Equation 1.7}$$

The resistance of the electrode with respect to remote earth is obtained dividing Equation 1.7 by the current $2I(0)$:

$$R = \frac{\rho}{2\pi}\cdot\frac{1}{L}\cdot\left[\ln\left[\frac{L}{a}\cdot\left[\sqrt{1+\left(\frac{a}{L}\right)^2} + 1\right]\right] + \frac{a}{L} - \sqrt{1+\left(\frac{a}{L}\right)^2}\right] \qquad \text{Equation 1.8}$$

And, in the case of $L>>a$, which is common in cathodic protection design, you can take the process to the limit when a/L approaches zero, obtaining the following.

$$R = \frac{\rho}{2\pi\cdot L}\cdot\left(\ln\left(2\cdot\frac{L}{a}\right) - 1\right) \qquad \text{Equation 1.9}$$

The Equations 1.8 or 1.9 are recommended when an electrode (anode) is very deep in the earth. Deep is defined when the distance between the top of the electrode (i.e., anode) and the soil surface is much larger than the anode's length.

If the electrode is closed to the surface of the earth, it induces charges underneath of it and, a different analysis is required, utilizing the ***method of images*** which will be developed in 1.3 below.

1.3 Potential and resistance calculation for the case of a vertical electrode with its top at the surface of the earth

Sunde provided equations to obtain the potential for an electrode flushed with the surface of the earth. In subsequent chapters, we will evaluate a slightly different approach, with a different coordinate system, that provides a better approximation of potential measurements on the surface of the earth, specifically for the case of cathodic protection and interference problems due to DC.

Although this second approach has a different analytical approach, the numerical results coincide with those obtained by Sunde's approximation, once they are corrected for the difference in coordinate systems.

This approach requires the use of the ***method of images***.

An electrical image is defined as:[5] "an electrified point or system of points on one side of a surface which would produce on the other side of that surface the same electrical action which the actual electrification of that surface really does produce". See Figure 1-3 below, where a charged point *1* below the surface ρ has image *1'* above the surface ρ. In this case, ρ represents the earth's surface, and δ is the distance between the charge and the point in the earth surface where the potential is calculated.

From Figure 1-3, it is noticeable that the current in the perpendicular direction to the surface ρ is zero.

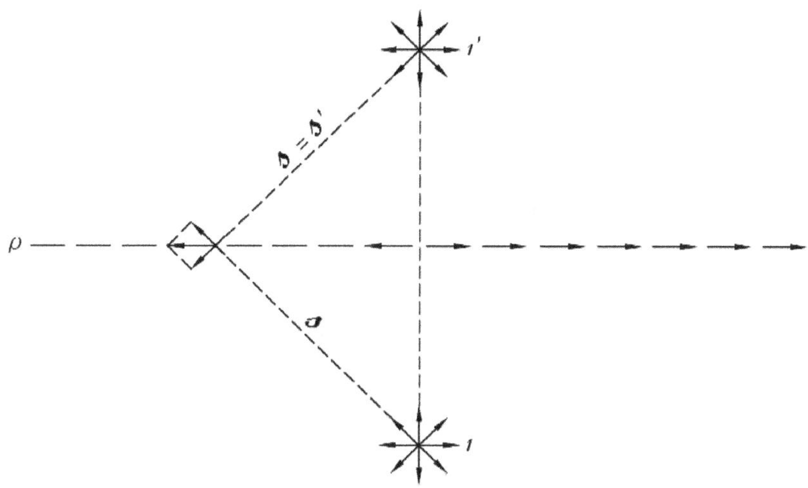

Figure 1-3: Electrical images.[6]

For the case of vertical electrodes, the image method can be illustrated as follows:

As a surface crosses the electrode at its midpoint, perpendicular to the electrode's axis and current is being injected into the electrode at the mid-point, both halves of the electrode are electrically in parallel. We can say then, that the potential of the surface bisecting the electrode in two equal sides is fixed. (Figure 1-4)

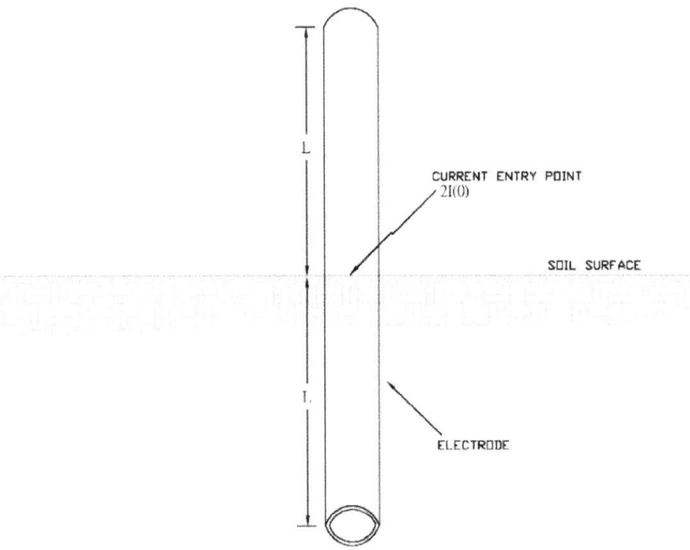

Figure 1-4: Electrical image method applied to a buried electrode.

Hence, if we remove the imaginary portion of the electrode (that segment above the surface of the earth) and its current *I(0)*, the potential and the current distribution in the lower part of the electrode will not change. Now, the resultant current in the electrode is not *2I(0)*, but just *I(0)*. So, to have the half of the current and not change the potential, the resistance of the lower part of the electrode must be the double of the total resistance of the entire electrode if the image is considered.

To obtain Equation 1.6, we integrated from *0* to *L/2*, to obtain the APM of the electrode. Now suppose that the half buried of the electrode has the length of *L* instead of *L/2*. And when you apply the APM you must divide by *L* instead of *L/2*.

$$V = \frac{\rho}{4\cdot\pi}\cdot\frac{I(0)}{L}\cdot 2L\cdot\left[\ln\left[\sqrt{\left(\frac{2L}{a}\right)^2 + 1} + \frac{2L}{a}\right] + \frac{a}{2L} - \sqrt{1 + \left(\frac{a}{2L}\right)^2}\right] \qquad \text{Equation 1.10}$$

We now input *I(0)* since we are only considering just one side of the electrode, and with length *L* instead of *L/2*.

Dividing by *L* and rearranging, we obtain the average potential over the entire electrode of length *L*.

$$V = \frac{\rho}{2\cdot\pi}\cdot\frac{I(0)}{L}\cdot\left[\ln\left[\frac{2L}{a}\cdot\sqrt{1 + \left(\frac{a}{2L}\right)^2} + 1\right] + \frac{a}{2L} - \sqrt{1 + \left(\frac{a}{2L}\right)^2}\right]$$

If we divide it by *I(0)*, we obtain the resistance of this electrode with respect to remote earth.

$$R = \frac{\rho}{2\cdot\pi}\cdot\frac{1}{L}\cdot\left[\ln\left[\frac{2\cdot L}{a}\cdot\left[\sqrt{1+\left(\frac{a}{2\cdot L}\right)^2}+1\right]\right]+\frac{a}{2\cdot L}-\sqrt{1+\left(\frac{a}{2\cdot L}\right)^2}\right] \quad \text{Equation 1.11}$$

If again, $L >> a$, the resistance of a vertical electrode with one end at the surface of the earth and receiving the current through this end, reduces to Equation 1.12.

$$R = \frac{\rho}{2\cdot\pi}\cdot\frac{1}{L}\cdot\left(\ln\left(\frac{4\cdot L}{a}\right)-1\right) \quad \text{Equation 1.12}$$

Examples of this are (i.e., flushed mounted vertical grounding rods used as AC grounding and mitigation systems for pipelines and equipment grounding):

The Equations 1.9 and 1.12 provide the resistance in the extreme cases where the electrode is infinitely deep or at the surface of the earth. We will also evaluate the case with the electrode at a finite distance below grade, which simulates an actual scenario used for surface anode cathodic protection design in the next section 1.4

1.4 Potential on the surface of the earth due to a vertical electrode

In the preceding sections, we evaluated the potential of a single electrode under two conditions, 1) in the space of infinite extent (remote earth) and 2) when the electrode is placed vertically, with its top touching the surface of the earth. In those cases, the Average Potential Method (APM) was used to calculate the influence of the charge in the axis of the electrode over the electrode surface, to obtain the resistance of the electrode respect to remote earth.

This approach is slightly different from the one used before, in Section 1.1. Here, we will use three dimensions coordinate system and the origin be on the surface of the earth.

In the case of cathodic protection, electrodes (anodes) are not placed on the surface of the earth but buried vertically or horizontally at a finite distance. Hence to simulate the potential at the surface of the earth due to a single cathodic protection anode, or group of anodes (groundbed) equations will be developed to provide more accurate estimates of the actual potentials at the surface of the earth, when considering the distance from the top of the anodes to the earth surface.

1.4.1 Potential on the surface of the earth due to a vertical electrode installed flush mounted at the surface of the earth, using a three-dimension coordinate system.

The x and y axes will represent a plane on the surface of the earth and the *z*-axis will be placed in the electrode's axis, with $z=0$ on the surface of the earth. This system of coordinates can be changed easily to the two-coordinate system using:

$$r = \sqrt{x^2 + y^2}$$

As a note, in the case of Sunde's work he used the variable *y* to represent the distance from the axis of the electrode to a point in the space. We are using the variable *r*. See Figure 1-5

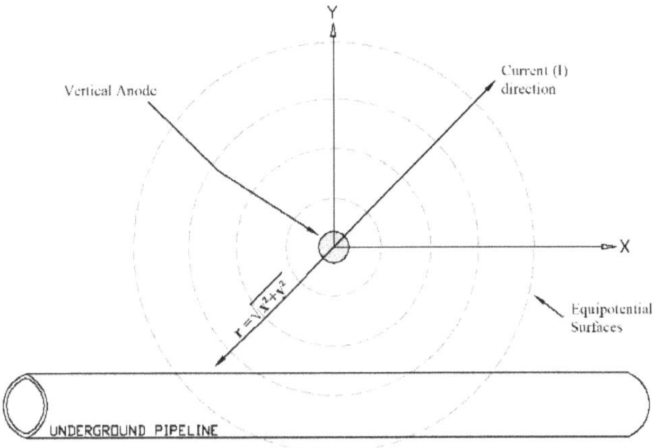

Note: Z axis is at the anode's center, where z = 0 is soil surface.

Figure 1-5: Potential in the surface of the earth due to a vertical buried electrode.

The distribution of the current around the electrode will be radial in nature and the effects of the electrode ends are not considered. The current density at a distance $(x,y,z=0)$ due to the current ***I*** leaving the electrode will be the total current divided by the area of a cylinder's surface with the following dimensions

The coordinate of a point over the electrode axis is: $x = 0 \quad y = 0 \quad z = z$

The coordinate of a point on the surface of the earth is: $x = x \quad y = y \quad z = 0$

The distance between the point in the electrode's axis and the point where the potential is determined is the radius of the cylinder forming equipotential lines moving away from the electrode's point. This is the square root that appears in Equation 1.13 below.

The surface's area is:

$$4\cdot\pi\cdot L\cdot\sqrt{(x-0)^2+(y-0)^2+(0-z)^2} \qquad \text{Equation 1.13}$$

The current density at the point *(x,y,0)*, due to the electrode element *dz* will then be:

$$\delta I = \frac{2I}{4\cdot\pi\cdot L\cdot\sqrt{(x-0)^2+(y-u)^2+(t)^2}}$$

or

$$\delta I = \frac{I}{2\cdot\pi\cdot L\cdot\sqrt{(x-0)^2+(y-0)^2+(0-z)^2}} \qquad \text{Equation 1.14}$$

The value of *2I* and *2L* were used because we are also considering the image of the electrode above the surface of the earth.

If ρ is the average soil resistivity in the area where the electrode is placed, the potential in the ground in the radial direction at a distance specified above, due to the current of an electrode element *dz* will be:

$$dV(x,y) = \frac{I\cdot\rho\cdot dz}{2\cdot\pi\cdot L\cdot\sqrt{(x-0)^2+(y-0)^2+(0-z)^2}} \qquad \text{Equation 1.15}$$

The total influence of the electrode over a point *(x,y,0)* will be the integral of Equation 1.15 over the entire length of the electrode:

$$V(x,y) = \frac{I}{2\cdot\pi}\cdot\frac{\rho}{L}\cdot\int_{-L}^{0}\frac{1}{\sqrt{(x)^2+(y)^2+(z)^2}}\,dz$$

The integral runs from *–L* to *0* to correspond with the origin of coordinates established at the beginning of this section, at the surface of the earth.

Because the integration is over *z*, we can define *r* as:

$$r = \sqrt{x^2+y^2}$$

The integral then reduces to

$$V(r) = \frac{I}{2\cdot\pi}\cdot\frac{\rho}{L}\cdot\int_{-L}^{0}\frac{1}{\sqrt{r^2+(z)^2}}\,dz$$

Using a substitution of variables, as we did previously:

$$\theta = \frac{z}{r} \quad d\theta = \frac{dz}{r}$$

$$z = -L \quad \theta = \frac{-L}{r} \qquad z = 0 \quad \theta := 0$$

The equation becomes:

$$V(r) = \frac{I}{2\cdot\pi}\cdot\frac{\rho}{L}\cdot\int_{\frac{-L}{r}}^{0} \frac{1}{\sqrt{1+\theta^2}}\,d\theta$$

The integral solution before evaluating the integration limits is then:

$$V(\theta) = \frac{I}{2\cdot\pi}\cdot\frac{\rho}{L}\cdot\ln\left(\theta + \sqrt{1+\theta^2}\right)$$

Now evaluating between $-L/r$ and 0, provides the following result:

$$V(x,y) = \frac{I}{2\pi}\cdot\frac{\rho}{L}\cdot\ln\left(\frac{\sqrt{L^2 + x^2 + y^2} - L}{\sqrt{x^2 + y^2}}\right) \qquad \text{Equation 1.16}$$

If we compare Equation 1.16 with that of Sunde's (Equation 3.54), where he uses y and we use, $\sqrt{x^2 + y^2}$, Equation 1.16 provides a greater accuracy when using the two variables analysis, when multiple anodes are used in the groundbed configuration. This will be developed further in Chapter 2. Otherwise, r can be used instead of $\sqrt{x^2 + y^2}$ when working with a single anode.

1.4.2 Potential on the surface of the earth due to a vertical electrode with the top of it on the surface of the earth as obtained by Rudenberg.

The equation known as Rudenberg equation[7], is equivalent to the Equation 1.16 above, with the main difference being the coordinate system chosen for the integral calculation.

The potential will be determined on the surface of the earth, due to the action of the electrode, where x-axis is placed in the axis of the electrode, with the origin in the middle of it. The origin coincides with the surface of the earth. The y-axis is perpendicular to the electrode axis and will be just at the ground level. The y-axis is radial starting at the electrode axis. See Figures 1-6a and 1-6b.

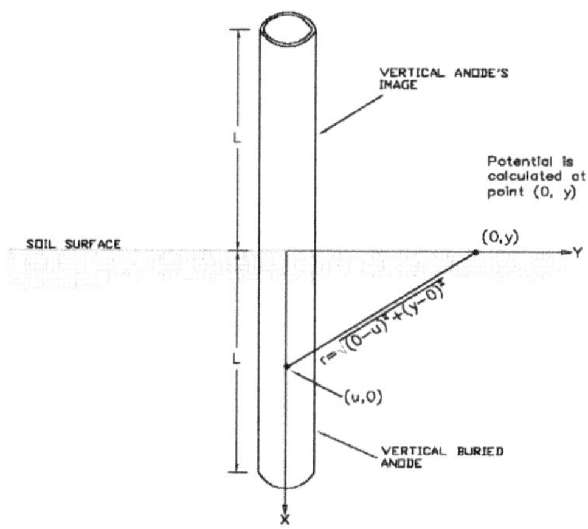

Figure 1-6a: Vertical electrode potential to remote earth in the surface of the ground. Elevation view

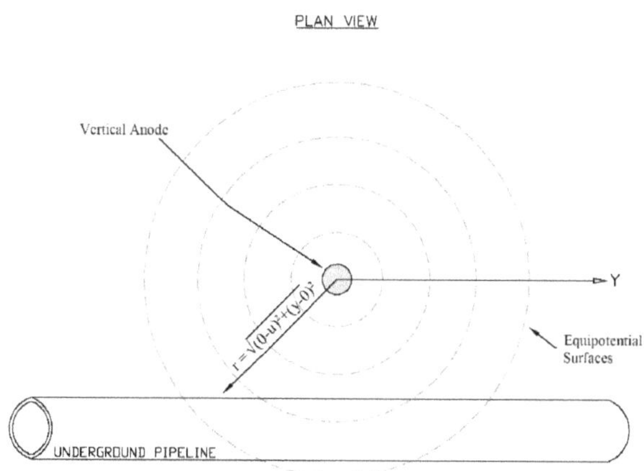

Figure 1-6b: Vertical electrode potential to remote earth in the surface of the ground, two axes approximation. Plain view

The coordinate of the point on the electrode axis is $x=u, y=0$, the coordinate of the point where the potential is desired is $x=0, y=y$. Like the approximation at the beginning of the chapter, y represents radial distance from the axis of the electrode.

The distance between them is: $r = \sqrt{(0-u)^2 + (y-0)^2}$

The differential potential is:

$$dV(y) = \frac{dI(u)}{du} \cdot \frac{\rho}{2 \cdot \pi} \cdot \frac{1}{\sqrt{u^2 + y^2}} \cdot du$$

The integral is as in the previous examples:

$$V(y) = \frac{\rho}{2 \cdot \pi} \cdot \frac{I(0)}{L} \cdot \int_0^L \frac{1}{\sqrt{u^2 + y^2}} \, du$$

Since the *u* variable runs in the electrode axis, it is always positive. Another way to see this is that because the image method is applied, the integral can be calculated on the positive side of the *x*-axis without impacting accuracy with respect to the absolute value of the potential at $x=0, y=y$.

Because the integral has the same form as previous calculations, we will show the result, and the reader can perform the necessary steps as shown in the previous developed equations.

Following integration, we obtain:

$$V(y) = \frac{\rho}{2 \cdot \pi} \cdot \frac{I(0)}{L} \cdot \left[\ln\left[\frac{L}{y} + \sqrt{1 + \left(\frac{L}{y}\right)^2} \right] - \ln\left(0 + \sqrt{1 + 0}\right) \right]$$

rearranging and using logarithm properties we find:

$$V(y) = \frac{\rho}{2 \cdot \pi} \cdot \frac{I(0)}{L} \cdot \ln\left(\frac{L + \sqrt{L^2 + y^2}}{y} \right)$$ Equation 1.17

That again is Equation 3.54 from Sunde's book[8], also known as Rudenberg equation, as it appears in Peabody's book[9]. The differences between Equation 1.17 and the ones mentioned above is that we use as total length of the electrode and its image $2L$ and Sunde used L. Peabody's shown y as our L and x as our y.

It is important to state that this equation is used to calculate potentials on the surface of the earth, due to the action of vertical electrodes that are flushed mounted. Here, the initial condition $x=0$ for the point where the potential is calculated is considered. It cannot be used to calculate potentials of points below grade $(x<0)$.

1.4.3 Case for three variable coordinate system with variable depth from the point where the potential is determined.

The case for determining the potential at points below ground, where $z<0$ is developed in the following Section 1.4.3, with three explicit variables: x, y and z. Now z-axis will represent the vertical electrode axis.

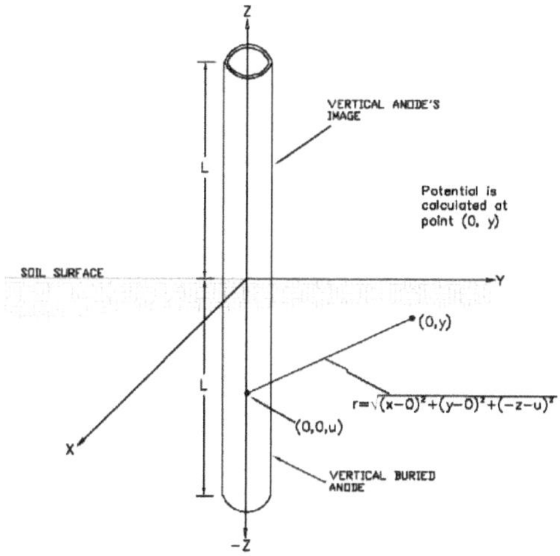

Figure 1-7: Potential due to Vertical electrode measured underground.

The coordinate of a point over the electrode axis is: $x = 0 \quad y = 0 \quad z = u$

The coordinate of a point at any depth in the earth, where the influence of the electrode (anode) is determined, is: $x = x \quad y = y \quad z = -z$

Then:

$$2\cdot\pi\cdot L\cdot\sqrt{(x-0)^2+(y-0)^2+(-z-u)^2}$$

The current density at the point $(x,y,-z)$, due to the electrode element du will then be:

$$\delta I = \frac{I}{2\cdot\pi\cdot L\cdot\sqrt{(x-0)^2+(y-0)^2+(-z-u)^2}}$$

The differential potential is:

$$dV(x,y,z) = \frac{I\cdot\rho\cdot du}{2\cdot\pi\cdot L\cdot\sqrt{(x-0)^2+(y-0)^2+(-z-u)^2}} \qquad \text{Equation 1.18}$$

And, as before, the total influence of the electrode over the point *(x,y,-z)* will be the integral of Equation 1.18, in the entire length of the electrode, but now with the variable of integration *u*:

And we arrive at:

$$V(x,y,z) = \frac{I}{2\pi}\cdot\frac{\rho}{L}\cdot\ln\left[\frac{L - z + \sqrt{x^2 + y^2 + (L-z)^2}}{-z + \sqrt{x^2 + y^2 + z^2}}\right] \qquad \text{Equation 1.19}$$

The reader can follow the integration process used before to arrive at the Equation 1.19.

Now, if we substitute the following.

$$y = \sqrt{x^2 + y^2}$$

and $z=0$, we arrived at the same Equation 1.17

$$V(y) = \frac{I}{2\pi}\cdot\frac{\rho}{L}\cdot\ln\left(\frac{\sqrt{L^2 + y^2} + L}{y}\right)$$

In this way, with Equation 1.19 we obtained a general equation that contains Equation 1.17 as a case when the potential is needed at the surface of the earth.

REFERENCES

1. E. D. Sunde, *Earth Conduction Effects in Transmission Systems*, 2nd ed. (New York, NY: Dover Publications, Inc., 1968), p. 69.
2. E. D. Sunde, *Earth Conduction Effects in Transmission Systems,* 2nd ed. (New York, NY: Dover Publications, Inc., 1968), p. 68.
3. O. D. Kellog, *Foundations of Potential Theory,* 2nd ed. (New York, NY: Dover Publications, Inc., 1954), p. 4.
4. W. Von Baeckmann, W. Schwenk, W. Prinz, eds., *Handbook of Cathodic Corrosion Protection: Theory and Practice of Electrochemical Protection Processes*, 3rd ed. (Houston, TX: Gulf Publishing Company), p. 536.
5. J. C. Maxwell, *A treatise in Electricity and Magnetism,* 3rd ed. Volume 1, (New York, NY: Dover Publications, Inc., 1954) p. 246, art. 157.
6. E. D. Sunde, *Earth Conduction Effects in Transmission Systems*, 2nd ed. (New York, NY: Dover Publications, Inc., 1968), paragraph 2.4.
7. A.W. Peabody, R.L. Bianchetti, ed., *Peabody's Control of Pipeline Corrosion,* 2nd ed. (Houston, TX: NACE International, 2000), p. 30.
8. E. D. Sunde, *Earth Conduction Effects in Transmission Systems*, 2nd ed. (New York, NY: Dover Publications, Inc., 1968), p. 91.
9. A.W. Peabody, R.L. Bianchetti, ed., *Peabody's Control of Pipeline Corrosion,* 2nd ed. (Houston, TX: NACE International, 2000), p. 30.

Potential and Resistance Equations for Vertical and Horizontal Electrodes When the Top of the Anode is at Depth t≠0

In this chapter, a three-dimensional coordinate system will be incorporated with the origin on the surface of the earth. The origin of the coordinate system will *not* coincide with the electrode's axis. Hence, the calculated potential values can be more accurately compared with the field-measured potential values that are measured from a single anode or group of anodes.

In the case of the mathematical method used by Sunde, discussed in Chapter 1, the axes of coordinates always coincide with the electrode's position on the top of the earth.

Typically, for cathodic protection, electrodes (anodes) are not placed at the surface of the earth, but buried vertically or horizontally, a finite distance below the surface. This chapter will provide equations that offer better estimates of the potential on the surface of the earth and resistance to remote earth, considering the distance from the top of the buried anodes to the earth surface.

It is important to mention that the use of a three-axes coordinate system permits separating the radial x and y axes in a planar component, which improves the calculation of ground-bed influence. These concepts will be further developed in Chapter 3.

2.1 Application of the method of images to determine electrode potential and resistance equations

An electrical image is defined as: "an electrified point or system of points on one side of a surface which would produce on the other side of that surface the same electrical action which the actual electrification of that surface really does produce."[1]

In the case of the buried electrode, like cathodic protection anodes, an electrical image will be an imaginary electrode with the same characteristics of the original electrode. The ground surface is the axis of symmetry.

The use of the image method avoids the need to calculate potentials and currents distribution within the confines of the semi space, where the actual electrode is bounded, in this case the confinement is the surface of the earth.

When determining the potential and resistance of buried electrodes, over their surface, the electrode and its image are in parallel. Consequently, to develop equations for determining the potential due to the electrode and its image, one can disregard the image electrode, together with its half space, and the potential and current profile will not change.

In this case, if the potential is kept the same, the amount of current will be one half, and the resistance of the electrode will be double the resistance in the case when the electrode and its image are considered together. It is basically the previously developed mathematical calculations provided in Chapter 1 for the resistance of the electrode when it is either near or at the surface of the earth.

2.2 Potential at surface of the earth due to a vertical electrode with its top at the depth t≠0

In the case of CP design, anodes are placed at a finite distance below the surface of the earth.

The case of the vertical electrode with its top at a depth $t \neq 0$ is discussed in Chapter 24 of the Von Baeckmann, Schwenk, and Prinz book.[2]

With that under consideration, potential equations will now be developed for the case of the vertical electrode when the top of it is at depth $t \neq 0$ and its image located with its bottom at t, above the surface of the earth (Figure 2-1). As can be seen, the main hypotheses will be the same found in Dwight's and Sunde's work that were explained in Chapter 1. Also, the integral found here has the same form that the one found in Chapter 1, Section 1.1, Equation 1.3.

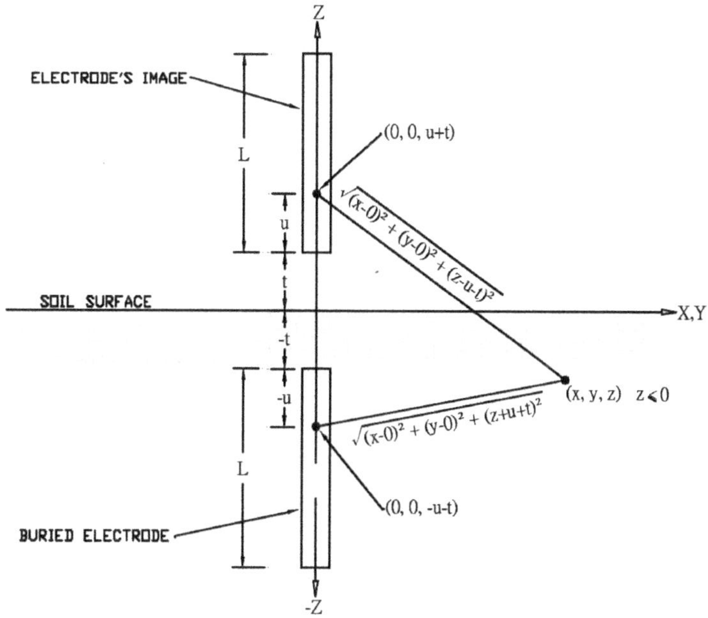

Figure 2-1: Vertical electrode at depth t > zero.

The coordinate of a point over the electrode axis is:

$$x = 0 \qquad y = 0 \qquad z = -u - t$$

where *t* is the depth of the top of the electrode. The coordinate of a point over the electrode's image axis is:

$$x = 0 \qquad y = 0 \qquad z = +u + t$$

t is the distance between the image's bottom and the surface of the earth. The coordinate of a point in the semi space (underground) is:

$$x = x \qquad y = y \qquad z = z \qquad z \leq 0$$

The distance between a point in the electrode and (x, y, z) in the semi space is:

$$\sqrt{(x-0)^2 + (y-0)^2 + (z+u+t)^2}$$

The distance between a point in the electrode's image and (x, y, z) is:

$$\sqrt{(x-0)^2 + (y-0)^2 + (z-u-t)^2}$$

The integral to obtain the potential will then be:

$$V(x,y,z) = \frac{\rho}{4\cdot\pi}\cdot\frac{I}{L}\cdot\int_0^L \left[\frac{1}{\sqrt{x^2+y^2+(z+u+t)^2}} + \frac{1}{\sqrt{x^2+y^2+(z-u-t)^2}}\right] du \quad \text{Equation 2.1}$$

Like before, the linear current density is *I(0)/L.*

For simplicity during the calculation, we will call $r = \sqrt{x^2+y^2}$

$$V(r,z) = \frac{\rho}{4\cdot\pi}\cdot\frac{I}{L}\cdot\int_0^L \left[\frac{1}{\sqrt{r^2+(z+u+t)^2}} + \frac{1}{\sqrt{r^2+(z-u-t)^2}}\right] du$$

$$V(r,z) = \frac{\rho}{4\cdot\pi}\cdot\frac{I}{L}\cdot\int_0^L \frac{1}{r}\cdot\left[\frac{1}{\sqrt{1+\left(\frac{z+u+t}{r}\right)^2}} + \frac{1}{\sqrt{1+\left(\frac{z-u-t}{r}\right)^2}}\right] du \quad \text{Equation 2.2}$$

And following the same procedure for this integral that the one used in Chapter 1:

$$V(r,z) = \frac{\rho}{4\cdot\pi}\cdot\frac{I}{L}\left[\ln\left[\frac{L+t+z+\sqrt{r^2+(z+L+t)^2}}{z+t+\sqrt{r^2+(z+t)^2}}\right] + \ln\left[\frac{z-t+\sqrt{r^2+(z-t)^2}}{z-t-L+\sqrt{r^2+(z-t-L)^2}}\right]\right] \quad \text{Equation 2.3}$$

Now, the potential at a point on the surface of the earth, where z=0, is:

$$V(r) = \frac{\rho}{4\cdot\pi}\cdot\frac{I}{L}\cdot\left[\ln\left[\frac{L+t+\sqrt{r^2+(L+t)^2}}{t+\sqrt{r^2+(t)^2}}\right] + \ln\left[\frac{-t+\sqrt{r^2+t^2}}{-t-L+\sqrt{r^2+(t+L)^2}}\right]\right] \quad \text{Equation 2.4}$$

This potential equation for a vertical electrode with its top a finite distance below grade and the potential is calculated at the surface of the ground. The equation is the same that appears in Table 24-1 Equation 7 presented by Von Baeckmann, et al.[3]

$$V1 = \frac{\rho}{2\cdot\pi}\cdot\frac{I}{L}\cdot\ln\left[\frac{L+t+\sqrt{x^2+y^2+(L+t)^2}}{t+\sqrt{x^2+y^2+t^2}}\right] \quad \text{Equation 2.5}$$

The second natural logarithm term in the Equation 2.4 is mathematically the same that the first, and can be rearranged to obtain Equation 2.5 and the variables are defined as follow:

L= length of the electrode and its image (m)
t= depth of the electrode (m)
ρ= soil resistivity near the buried electrode (ohm×m)
I= current injected at the electrode (A)
x, *y*, and *z* define the position variables.

2.3 Resistance to remote earth of a vertical electrode with the top at depth t≠0.

Computing the average potential at the anode surface, with radius a, diameter *d*, depth to the top of the anode where $t<0$.

To be consistent with the axes of coordinates used above in calculating the potential due to the electrode and its image, the integration limits will be defined as $(-L-t)$, as the lower limit and $(-t)$ as the upper limit.

Since the electrode is a finite distance below the surface of the earth, Equation 2.3 is used to determine the potential at a point on the surface of the anode. The current leaving the image of the electrode will not impact the calculations so, one can disregard the electrode's image, without affecting the potential or the current of the electrode. Once again, to maintain a constant potential the electrode resistance will double.[4]

Applying the average potential method, use the following steps:

- Evaluate Equation 2.3 for the case of $r=a$. To obtain the potential at a point on the surface of the anode.
- Integrate the Equation 2.3 between $-L-t$ and $-t$. To obtain the total potential at the surface of the anode.
- Divide the result by L, the length of the electrode. To obtain the average potential due to all the electrode charges.
- Divide by I, the current of the electrode. To obtain the resistance of the electrode to remote earth.

Following the steps mentioned above:

Step 1: Evaluating Equation 2.3 with $r=a$ provides the following:

$$V(a,z) = \frac{\rho}{4\cdot\pi}\cdot\frac{I}{L}\cdot\left[\ln\left[\frac{L+t+z+\sqrt{a^2+(z+L+t)^2}}{z+t+\sqrt{a^2+(z+t)^2}}\right]+\ln\left[\frac{z-t+\sqrt{a^2+(z-t)^2}}{z-t-L+\sqrt{a^2+(z-t-L)^2}}\right]\right] \quad \text{Equation 2.6}$$

In Equation 2.6 substitute r for a**,** the electrode's radius. The radius r of the electrode coincides with the distance between the charges in its axis and the surface of the electrode, where the resistance to remote earth is calculated. See Step 2.

Step 2: Calculating the total potential at the surface of the electrode. *(-t to -(L+t))*

Now, to determine the average potential over the surface of the vertical electrode, the integral has the form:

$$Vav(a) = \frac{\rho}{4\cdot\pi}\cdot\frac{I}{L}\cdot\frac{1}{L}\cdot\int_{-(L+t)}^{-t}\left[\ln\left[\frac{L+t+z+\sqrt{a^2+(z+L+t)^2}}{z+t+\sqrt{a^2+(z+t)^2}}\right]+\ln\left[\frac{z-t+\sqrt{a^2+(z-t)^2}}{z-t-L+\sqrt{a^2+(z-t-L)^2}}\right]\right]dz$$

Using logarithm properties, this integral can be separated into four integrals. The four integrals are:

$$A = \int_{-(L+t)}^{-t} \left[\ln\left[L + t + z + \sqrt{a^2 + (z + L + t)^2} \right] \right] dz$$

$$B = \int_{-(L+t)}^{-t} \left[\ln\left[z + t + \sqrt{a^2 + (z + t)^2} \right] \right] dz$$

$$C = \int_{-(L+t)}^{-t} \left[\ln\left[z - t + \sqrt{a^2 + (z - t)^2} \right] \right] dz$$

$$D = \int_{-(L+t)}^{-t} \left[\ln\left[z - t - L + \sqrt{a^2 + (z - t - L)^2} \right] \right] dz$$

$$Vav(a) = \frac{\rho}{4 \cdot \pi} \cdot \frac{I}{L} \cdot \frac{1}{L} \cdot (A - B + C - D)$$

The solution for the electrode resistance after integration, and completing the steps mentioned above is:

$$Rr = \frac{Vav(a)}{I}$$

$$Rr1 = \frac{\rho}{4\pi} \cdot \frac{1}{L} \cdot \left[\ln\left(\frac{\sqrt{L^2 + a^2} + L}{\sqrt{L^2 + a^2} - L} \right) + 2 \frac{(L + t)}{L} \cdot \ln\left[\frac{\sqrt{(L + 2t)^2 + a^2} - (L + 2t)}{\sqrt{4 \cdot (L + t)^2 + a^2} - 2 \cdot (L + t)} \right] \right]$$

$$Rr2 = \frac{\rho}{4\pi} \cdot \frac{1}{L} \cdot \left[\frac{2 \cdot t}{L} \cdot \ln\left[\frac{\sqrt{(L + 2t)^2 + a^2} - (L + 2t)}{\sqrt{(2 \cdot t)^2 + a^2} - 2t} \right] \right]$$

$$Rr3 = \frac{\rho}{4\pi} \cdot \frac{1}{L} \cdot \left[\frac{2\sqrt{(L + 2t)^2 + a^2}}{L} - \frac{\sqrt{2t^2 + a^2}}{L} - \frac{\sqrt{4(L + t)^2 + a^2}}{L} - \frac{2\sqrt{L^2 + a^2}}{L} + \frac{2a}{L} \right]$$

$$Rr = Rr1 + Rr2 + Rr3$$

Equation 2.7

Equation 2.7 is a generalization of Sunde's Equation 3.24[5] for the vertical electrode. As previously described, Sunde's equation is a case when the depth of the vertical electrode is *t=0*.

Von Baeckmann and Schwenk also presented an equation for the case when the top of the anodes is below grade. The difference in values using the Von Baeckmann and Schwenk equation and the Equation 2.7 when *t=0* is about 3% and when *t≠0*, the difference is about 7%.

Also, when t approaches infinity, $t \geq 300$ft, Equation 2.7 corresponds to Sunde's Equation 3.08[6] for the case of an electrode in a medium of infinite extent in all directions, and Equation 1.9 in Chapter 1 of this book.

2.3.1 The case of potential due to a horizontal electrode (anode) at depth *t*, below the surface of the earth

The potential and resistance of horizontal electrodes, with its center on the surface of the earth, does not have any practical application for cathodic protection; most likely it is used for the case of shallow electrical grounding systems.

As in the case of the vertical electrode, the origin of coordinates is on the surface of the earth. To be consistent with other developments, we will place the y-axis as the electrode axis. One end of the electrode will coincide with the origin of $x=0, y=0$, and $z=-t$ where t is the distance from the surface of the earth to the axis of the electrode. The image of the electrode on the surface of the earth is also considered, as adding influence over the point where the potential is determined. This image is over the surface of the earth, at the distance t (Figure 2-2). So, the distance between the electrode's axis and its image's axis is $2t$. The current entering the electrode and its image is I.

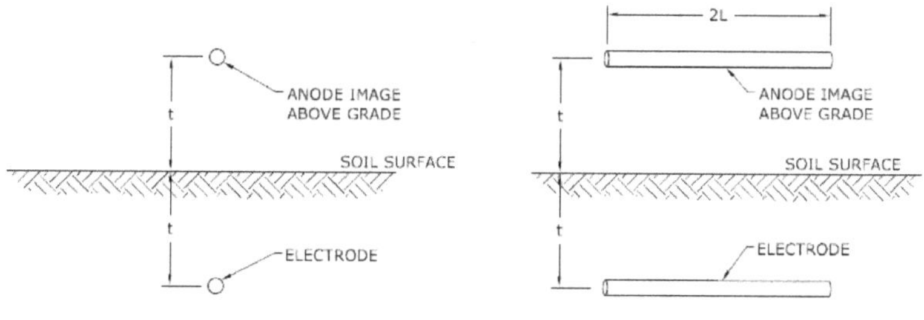

Figure 2-2: Horizontal electrode at depth *t* and its image at height *t*.

All the other hypotheses about the current distribution and current density that applied to vertical electrodes also apply here.

First, we will obtain the potential of the electrode and its image in a point on the surface of the earth.

The coordinate of a point over the electrode axis is: $x = 0 \quad y = u \quad z = -t$

The coordinate of a point over the electrode's image axis is: $x = 0 \quad y = u \quad z = t$

The coordinate of a point on the surface of the earth is: $x = x \quad y = y \quad z = 0$

The distance between a point on the electrode axis and a point on the surface of the earth is: $\sqrt{(x-0)^2+(y-u)^2+(0+t)^2}$. In the case of the electrode's image is:

$\sqrt{(x-0)^2+(y-u)^2+(0-t)^2}$.

The current density due to the electrode and the electrode's image will be determined over equipotential surfaces with a radius equal to the distances above, where $2L$ is the length of the horizontal electrode. In this way, the current density due to the electrode and its image is:

$$\delta I = \frac{I}{4\cdot\pi\cdot L}\cdot\frac{1}{\sqrt{(x-0)^2+(y-u)^2+(t)^2}}$$

Now if ρ is the soil resistivity, the differential potential in the ground at a distance specified above, due to an electrode element *du* is:

$$dV(x,y,u) = \frac{I\cdot\rho\cdot du}{4\cdot\pi\cdot L\cdot\sqrt{(x)^2+(y-u)^2+(t)^2}} \qquad \text{Equation 2.8}$$

If the electrode is located with the electrode's axis on the surface of the earth, where $t=0$, then Equation 2.8 will be:

$$dV(x,y,u) = \frac{I\cdot\rho\cdot du}{4\cdot\pi\cdot L\cdot\sqrt{x^2+(y-u)^2}} \qquad \text{Equation 2.9}$$

With the electrode at depth t, the total influence of the electrode and its image over the point $(x,y,0)$ is the integral of Equation 2.8, over the entire length of the electrode $2L$, see Figure 2-2:

$$V(x,y) = \frac{I}{4\cdot\pi}\cdot\frac{\rho}{L}\cdot\int_0^{2L}\frac{1}{\sqrt{x^2+(y-u)^2+(t)^2}}\,du$$

The integral runs from 0 to $2L$ to correspond with the length of the electrode per Figure 2-2. The integration variable u is along the electrode axis (y), as in the cases before, when we calculated the vertical electrode potential.

For simplicity, during integration, we will define $r=\sqrt{x^2+t^2}$

$$V(r,y) = \frac{I}{4\cdot\pi}\cdot\frac{\rho}{L}\cdot\int_0^{2L}\frac{1}{\sqrt{(r)^2+(y-u)^2}}\,du$$

Following the same procedure for integration as developed in Chapter 1 we have,

$$V(r,y) = \frac{I}{4\cdot\pi}\cdot\frac{\rho}{L}\cdot\int_{\frac{y-2L}{r}}^{\frac{y}{r}} \frac{1}{\sqrt{1+\zeta^2}}\,d\zeta$$

The integrals have the following solution without integration limits:

$$V(\zeta) = \frac{I}{4\cdot\pi}\cdot\frac{\rho}{L}\cdot\ln\left(\zeta + \sqrt{1+\zeta^2}\right)$$

Evaluating the integral:

$$V(r,y) = \frac{I}{4\pi}\cdot\frac{\rho}{L}\cdot\left[\ln\left[\frac{y}{r} + \sqrt{1+\left(\frac{y}{r}\right)^2}\right] - \ln\left[\frac{y-2L}{r} + \sqrt{1+\left(\frac{y-2L}{r}\right)^2}\right]\right]$$

Using logarithm properties, and substituting *r* by its value:

$$V(x,y) = -\frac{I}{4\pi}\cdot\frac{\rho}{L}\cdot\left[\ln\left[\frac{y+\sqrt{x^2+y^2+t^2}}{y-2L+\sqrt{x^2+(y-2L)^2+t^2}}\right]\right] \qquad \text{Equation 2.10}$$

If the length of the electrode is *L* instead of *2L*, the equation will then be:

$$V(x,y) = -\frac{I}{2\pi}\cdot\frac{\rho}{L}\cdot\left[\ln\left[\frac{y+\sqrt{x^2+y^2+t^2}}{y-L+\sqrt{x^2+(y-L)^2+t^2}}\right]\right] \qquad \text{Equation 2.11}$$

If we go a step further and generalize Equation 2.11 for the case of the potential calculated in any point that could be in the surface or deep in the ground, Equation 2.11 is then, with $z < 0$, and considering the image, then Eq. 2.11 can be as follows:

$$V(x,y,z) = \frac{I}{4\pi}\cdot\frac{\rho}{L}\cdot\left[\ln\left[\frac{y+\sqrt{x^2+y^2+(z+t)^2}}{y-L+\sqrt{x^2+(y-L)^2+(z+t)^2}}\right] + \ln\left[\frac{y+\sqrt{x^2+y^2+(z-t)^2}}{y-L+\sqrt{x^2+(y-L)^2+(z-t)^2}}\right]\right] \qquad \text{Equation 2.11.1}$$

Equation 2.11 is equivalent to Sunde's Equation 3.55,[7] which assumes *x* is the electrode axis and the current and the origin of coordinates are at the center of the electrode.

Applying Equation 2.11 and comparing with Sunde's Equation 3.55, using these inputs:

$$I = 2 \text{ amp} \quad \rho = 1000 \text{ ohm}\cdot\text{cm} \quad L = 1 \text{ m} \quad t = 1 \text{ m}$$

We can evaluate Equation 2.11 as such:

$$V(2m, 0.5m) = 1.412V$$

Now we will evaluate Sunde's Equation 3.55, with the same input values of *x* and *y* adjusted, due to the change in the origin of the coordinate system. Remember that in the case of Sunde, the origin of coordinate coincides with the axis of the electrode for *y* and in the middle of the electrode with *x*. That equation and result are:

$$Vs(x,y) = \frac{I}{2\pi}\cdot\frac{\rho}{L}\cdot\left[\ln\left[\frac{x+\frac{L}{2}+\sqrt{\left(x+\frac{L}{2}\right)^2+y^2+t^2}}{x-\frac{L}{2}+\sqrt{\left(x-\frac{L}{2}\right)^2+y^2+t^2}}\right]\right]$$

When calculated the value of Vs is:

$$Vs(0m, 2m) = 1.412V$$

The difference between the two equations is due to the origin of coordinates, yet the results are equivalent.

Equation 2.11 provides a simpler formula for application to cathodic protection design.

Equation 2.11.1 can be considered more accurate and flexible than Equation 2.11. Equation 2.11.1 facilitates measuring the distance in a three-dimension system of coordinates, and the point where the potential is calculated would even be underground, while Equation 2.11 only allows to measure potentials radial distances from the axis of the electrode at the surface of the ground.

2.4 Obtain the resistance's equation for a horizontal electrode at depth *t*, following Dwight and Sunde's method

In the case of the resistance of a horizontal electrode at depth *t*:

- Consider the resistance of the electrode calculated from the potential at the surface of the electrode due to its charge, concentrated in its axis, and
- the resistance calculated from the potential due to the charge in the axis of the electrode's image.

The action of the potential developed by the image over the actual electrode increases the resistance of the electrode with respect to remote earth and must be considered.

Another consideration in the resistance calculation is that the electrode and its image are electrically in parallel; the resistance of the electrode and its image is half of the resistance of the electrode by itself.

The part of the resistance due to the electrode's charge is obtained using Equation 1.9 from Chapter 1.

$$R = \frac{\rho}{2\pi} \cdot \frac{1}{L} \cdot \left(\ln\left(\frac{2L}{a}\right) - 1 \right)$$

where L is the electrode's length.

Remember previously when the resistance was needed on the surface of the electrode due to the charge at its center, we used the distance a, the electrode's radius. Now the charge is in the image's axis, so using the same reasoning, but using a distance $2t$, the distance between the electrode axis and the image axis, and substituting in Equation 1.8, the equation for the influence of the electrode's image is calculated as follows:

$$Ri = \frac{\rho}{2\pi} \cdot \frac{1}{L} \cdot \left[\ln\left[\frac{L}{2 \cdot t} \left[\sqrt{1 + \left(\frac{2 \cdot t}{L}\right)^2} + 1 \right] \right] + \frac{2 \cdot t}{L} - \sqrt{1 + \left(\frac{2 \cdot t}{L}\right)^2} \right] \quad \text{Equation 2.12}$$

Now, rearranging Equation 2.12 in a more recognized form:

$$Ri = \frac{\rho}{2\pi} \cdot \frac{1}{L} \cdot \left[\ln\left[\frac{L + \sqrt{L^2 + (2 \cdot t)^2}}{2 \cdot t} \right] + \frac{2 \cdot t}{L} - \frac{\sqrt{(2 \cdot t)^2 + (L)^2}}{L} \right] \quad \text{Equation 2.13}$$

Dwight provides a very detailed explanation of the development of Equation 2.12.[8] In our case, the potential is determined not to be due to the charge of the electrode at its axis, but due to the charge of its image, concentrated in the image's axis.

Entering $2t$ instead of a in Equation 1.5, with all the other conditions being the same, you are calculating the average potential in the axis of the electrode due to the charge in the axis of its image and vice versa. Having to consider the image of the horizontal electrode on the resistance calculation explains why the resistance of horizontal electrodes near the surface of the ground is larger than the resistance of an electrode with the same size installed vertically.

The hypotheses behind these calculations are:

1. The distance between the electrode and its image $(2t >> a)$ is much larger than the radius of the electrode.
2. The radius of the electrode is much less than its length $(r << L)$, so that the charge can be considered as concentrated in its axis, instead of its surface.

Now, the total resistance of a horizontal electrode located closed to the surface of the earth, considering its image is the sum of Equations 1.9 and 2.13:

$$R = \frac{\rho}{2\pi} \cdot \frac{1}{L} \cdot \left[\ln\left(\frac{2L}{a}\right) - 1 + \left[\ln\left[\frac{L + \sqrt{L^2 + (2 \cdot t)^2}}{2 \cdot t} \right] + \frac{2 \cdot t}{L} - \frac{\sqrt{(2 \cdot t)^2 + L^2}}{L} \right] \right] \quad \text{Equation 2.14}$$

This is Equation 19 obtained by Dwight, who calculated for an electrode with length *2L*.

It is important to note here that in the case of Equation 2.14 we do not meet the limit of $2t<<L$ as applied by Sunde.[9] In the case of CP design, using regular anodes, the length of the anodes is typically less than twice the distance between the electrode axis and the surface of the earth (distance between the electrode and its image is $2t$). In the case where $L>>2t$, as is the case of several horizontal anodes in the same coke bed, strip sacrificial anodes, long continuous impressed current ceramic anodes, or AC mitigation grounding systems, the Equation 2.14 will always provide a much more accurate result.

2.5 Obtain the potential's equation for a horizontal electrode at depth *t*, where the pipeline and the anode are not parallel

The equations developed in Section 2.4 are for the cases where the horizontal anode (or group of horizontal anodes in the same coke backfill) are parallel or perpendicular to the protected structure or pipeline.

Equations can be developed evaluating anodes installed at an angle between the pipeline and the horizontal anode (Figure 2-3).

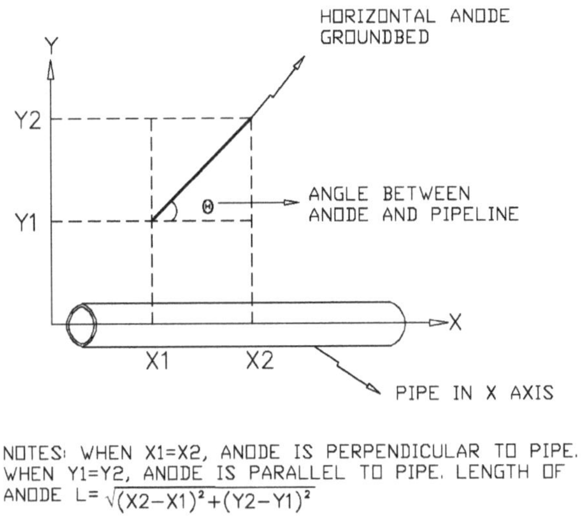

Figure 2-3: Horizontal anode (groundbed) placed any angle respect to pipeline (pipeline is in the x-axis).

These calculations will be developed utilizing the cartesian axes system, with the following assumptions.

The pipeline is located with its axis in the x-axis. The horizontal anode or linear groundbed starts at *(x1,y1)* and ends at *(x2,y2)*.

The length of the horizontal electrode is:

$$L = \sqrt{(x2 - x1)^2 + (y2 - y1)^2}$$

From here, we can deduce the necessary variables as follow:

$$\sin = \frac{y2 - y1}{L}$$

This is the *sine* of the angle formed between the pipeline (in the x-axis) and the groundbed.

$$\cos = \frac{x2 - x1}{L}$$

This is the *cosine* of the angle formed between the pipeline (in the *x*-axis) and the groundbed.

Based on the same principles that were explained in Chapter 1, the variable of integration *(u)* to obtain the potential equation will be at the groundbed axis.

The integral equation for this case will be:

$$Ug(x,y) = \frac{I}{2\pi} \cdot \frac{\rho}{L} \cdot \left[\int_0^L \left[\frac{1}{\sqrt{(x - x1 - u \cdot \cos)^2 + (y - y1 - u \cdot \sin)^2 + (t)^2}} \right] du \right] \quad \text{Equation 2.15;}$$

where:

- I is the current entering the groundbed
- ρ is the resistivity of the soil around the groundbed
- t is the depth from the center of the groundbed to soil surface

If the reader has a program like Mathcad or Matlab, they can use the integral equation and calculate the value of the potential directly from the Equation 2.15.

For those working in Excel or using a calculator, it may be more suitable to use Equation 2.16:

$$a(x,y) = \frac{\sqrt{L^2 - 2 \cdot L \cdot [(x - x1) \cdot \cos + (y - y1) \cdot \sin] + (x - x1)^2 + (y - y1)^2 + t^2}}{\left[\sqrt{(x - x1)^2 + (y - y1)^2 + t^2} - [(x - x1) \cdot \cos + (y - y1) \cdot \sin]\right]}$$

$$b(x,y) = \frac{L - [(x - x1) \cdot \cos + (y - y1) \cdot \sin]}{\left[\sqrt{(x - x1)^2 + (y - y1)^2 + t^2} - [(x - x1) \cdot \cos + (y - y1) \cdot \sin]\right]}$$

$$Uae1(x,y) = \frac{Ides}{2\pi} \cdot \frac{\rho}{L} \cdot \ln(a(x,y) + b(x,y)) \quad \text{Equation 2.16}$$

To compare this case of horizontal anode paralleling one of the axes, we will compare the calculations using Equations 2.15 and 2.16, with the calculations using Equation 2.11, that provided the potentials for the case of horizontal anode parallel to *y*-axis.

Equation 2.11 for the case when the anode is parallel to the pipeline at a distance *(x1,y1)*, from the origin of coordinates, is written as:

$$Vy(x,y) = \frac{I}{2\cdot\pi}\cdot\frac{\rho}{L}\cdot\ln\left[\frac{y - y1 + \sqrt{(x - x1)^2 + (y - y1)^2 + t^2}}{y - y1 - L + \sqrt{(x - x1)^2 + (y - y1 - L)^2 + t^2}}\right] \quad \text{Equation 2.17}$$

If the horizontal anode is parallel to *y*-axis, $x1=x2$.

The variables use below are shown in Figure 2-3.

For this example, the input variables are the following values:

$x1$ = 2 ft
$y1$ = 50 ft
$y2$ = 276 ft
L = 226 ft
ρ= 20,000 ohm•cm
I= 9 A

The potential value calculated with the Equations 2.15, 2.16, and 2.17 are the same = ***7.094 volt***.

This result demonstrates that Equation 2.17 represents a specific case of Equations 2.15 and 2.16.

REFERENCES

1. J. C. Maxwell, *A treatise in Electricity and Magnetism*, 3rd ed. Volume 1, (New York, NY: Dover Publications, Inc., 1954) p. 246, art. 157.
2. W. Von Baeckmann, W. Schwenk, W. Prinz, eds., *Handbook of Cathodic Corrosion Protection: Theory and Practice of Electrochemical Protection Processes*, 3rd ed. (Houston, TX: Gulf Publishing Company), p. 543.
3. W. Von Baeckmann, W. Schwenk, W. Prinz, eds., *Handbook of Cathodic Corrosion Protection: Theory and Practice of Electrochemical Protection Processes*, 3rd ed. (Houston, TX: Gulf Publishing Company), p. 539.
4. E. D. Sunde, *Earth Conduction Effects in Transmission Systems*, 2nd ed. (New York, NY: Dover Publications, Inc., 1968), p. 46.
5. E. D. Sunde, *Earth Conduction Effects in Transmission Systems*, 2nd ed. (New York, NY: Dover Publications, Inc., 1968), p. 75
6. E. D. Sunde, *Earth Conduction Effects in Transmission Systems*, 2nd ed. (New York, NY: Dover Publications, Inc., 1968), p. 70

7. E. D. Sunde, *Earth Conduction Effects in Transmission Systems*, 2nd ed. (New York, NY: Dover Publications, Inc., 1968), p. 90
8. H.B. Dwight, "Calculation of Resistances to Ground," in Classic Papers and Reviews on Anode Resistance Fundamentals and Applications, ed. R.L. Benedict, (Houston, TX: NACE, 186), p. 3.
9. E. D. Sunde, *Earth Conduction Effects in Transmission Systems*, 2nd ed. (New York, NY: Dover Publications, Inc., 1968), p. 80

Potential Equations for Multiple Vertical and Horizontal Electrodes in Various Groundbed Configurations

In addition to the theory and principles explained in previous chapters for single electrodes, this chapter will address the additive effect of multiple electrodes.

These equations have been developed to assist cathodic protection (CP) engineers when studying the influence of a groundbed on any buried metallic structure in the vicinity or around that groundbed. These equations can also be used:

- To determine and minimize the influence of stray DC prior to installation of a groundbed.
- To calculate the potential influence on an existing pipeline due to its own groundbed.

This chapter also introduces the remote earth concept as a key variable to be considered, as part of the potential calculations. We present a Mathcad subroutine using these equations for multiple anode configurations to calculate the remote earth distance.

3.1 Determination of remote earth (ξ)

The remote earth definition was addressed in Peabody's book as follows:

> *"Current discharge, from an anode or group of anodes forming the ground bed, will cause voltage drops in the earth between point along lines radiating from the ground bed. Close to the ground bed, the voltage drop per unit of distance is relatively high. As one moves away from the groundbed, this voltage drop per unit of distance becomes less and less until a point is reached beyond which no further significant voltage drop can be observed. This point may be considered as remote earth and establishes the radius of what is termed the area of influence surrounding the groundbed"*[1]

It is important to point out here that what Peabody calls "voltage drop per unit of distance" is the potential gradient.

The point where the potential gradient is zero, indicates no influence of the groundbed. Based on this, the percentage of the potential gradient from the groundbed can be effectively calculated. This percentage is determined by comparing the potential in the ground at ***the groundbed to the potential to remote earth of the groundbed***. The potential to remote earth is obtained by multiplying the groundbed's resistance to remote earth by the current applied to the groundbed.

The value of ξ (remote earth distance) is determined as a condition of what percentage of the groundbed influence is acceptable. Von Baeckmann[2] and NACE CP4[3] describe the importance of calculating the ratio of the potential in the ground to remote earth due to a groundbed and the potential of the groundbed to remote earth.

The next section shows that calculated difference between the potential in the ground due to the groundbed and the potential of the groundbed to the remote earth. Additionally, Section 3.2 provides procedures for performing field measurement to properly obtain remote earth.

As a practical rule of thumb, the maximum influence on an underground structure to consider to be remote from a groundbed is: the potential in the ground due to the groundbed as compare with the potential of the groundbed to remote earth is 5% or less. We use this percentage in the following calculations to determine remote earth.

A subprogram in Mathcad is shown in the following example to obtain remote distance.

In this example, we are using the equation 3.4.1 below (Section 3.3), for vertical anodes in a groundbed configuration where the groundbed is perpendicular to the pipeline.

The variables used in the following example are defined as follows:

Rprom: the resistance of the groundbed to remote earth
I: the current injected on the groundbed
ρ: the soil resistivity near the groundbed
N: the number of anodes that form the groundbed
L: the length of the active column where each anode is immersed
V_{an}: the potential ***of the*** groundbed to remote earth
V(x,y): the potential ***in the*** ground ***due to*** the groundbed
S: the distance between anodes in the groundbed configuration
t: the depth of the top of the coke active column
x and ***y*** are as defined in previous chapters and represent the coordinates for a point on the surface of the earth where the potential in the ground is calculated.

Remdist is the Mathcad subroutine.

$$\text{Rprom} = 2\cdot\text{ohm} \quad \text{I} = 3\text{amp} \quad \text{N} = 10 \quad \rho = 4000\text{ohm}\cdot\text{cm}$$

$$\text{Van} := \text{Rprom}\cdot\text{Ides} \quad \text{Van} = 6\cdot\text{volt}$$

$$\text{L} = 4\text{m} \quad \text{t} = 4\cdot\text{m} \quad \text{S} = 5\text{m}$$

Now substituting o, the common factor for all the anodes in a groundbed configuration, results in the following:

$$\text{o} = \frac{\text{I}\cdot\rho}{4\cdot\pi\,\text{N}\cdot\text{L}}$$

Note that the condition to obtain ξ is that the value of $V(0,0)$ be equal to or less than 0.05 times (5%) the potential of the groundbed to remote earth, V_{an}, when the groundbed is ξ feet away from the investigated structure in the y-axis direction. In this example the equation is used only to calculate the potential on the surface of the ground. The case for potentials at any depth underground will be discussed later.

When working with Mathcad programming the variables must be unitless, that is why the defined new variables inside the subroutine are unitless for the calculations to proceed.

In this example, the variable x is fixed and the subroutine to calculate the value of $V(x,y)$ is when y is increasing (that is, moving away from the protected structure)—which is located in the x-axis at the $y=0$ coordinate.

$$\text{Remdist} := \text{for } y \in 1..100000$$

$$\left|\begin{array}{l} x \leftarrow 0 \\ \text{S1} \leftarrow \dfrac{\text{S}}{\text{m}} \\ \text{Va1} \leftarrow \dfrac{\text{Van}}{\text{volt}} \\ \text{t1} \leftarrow \dfrac{\text{t}}{\text{m}} \\ \text{L1} \leftarrow \dfrac{\text{L}}{\text{m}} \\ \text{o1} \leftarrow \dfrac{\text{o}}{\text{volt}} \\ \text{V} \leftarrow \displaystyle\sum_{n=0}^{N-1}\left[2\text{o1}\cdot\ln\left[\frac{\text{t1} + \text{L1} + \sqrt{x^2 + (n\cdot\text{S1} - y)^2 + (\text{t1} + \text{L1})^2}}{\text{t1} + \sqrt{x^2 + (n\cdot\text{S1} - y)^2 + \text{t1}^2}}\right]\right] \\ \text{return } y \text{ if } \dfrac{\text{V}}{\text{Va1}} < 0.05 \end{array}\right.$$

$$\text{Remdist} = 90$$

Dist=Remdist m
Dist=90 m (~300 ft)

With the variables that were assigned, the above calculation shows that any buried metallic structure at a distance equal to or greater than 90 m (300 ft) from the groundbed in the *y*-axis direction is affected by 5% or less than the groundbed influence. If the groundbed influence wants to be determined in a direction other than *y*-axis, the calculation of remote earth condition in the subprogram above can be modified accordingly.

Due to the non-uniformity of the soils, the presence of other existing buried metallic structures, and another source of DC, the measurement of potential in the field is complicated, and sometimes not accurate.

As it is well explained in NACE CP4,[4] one of the advantages of calculating the desired degree of remoteness with the method provided above **does not depend** on the soil resistivity or the current values. Both appear in the calculation of the groundbed potential to remote earth, and in the equation for the calculation of the potential in the ground due to the groundbed. Comparing one value with the other cancels out the dependence for these two variables.

The ability to calculate the groundbed potential allows for:

1. The determination of interference, in the case of foreign structures,
2. Better selection of groundbed locations, and
3. IR drop calculation on the affected structure.

In the following section a procedure will be discussed on how to measure these potentials in the field and how to determine remote earth.

3.2 Field evaluation of remote earth

A methodology to obtain true remote earth from groundbeds and the structure being evaluated is presented below. This will apply when we need to determine remote earth from a tank farm, a pipeline, an anode, or a groundbed.

3.2.1 Remote earth distance measured using one reference Cell (Method 1)

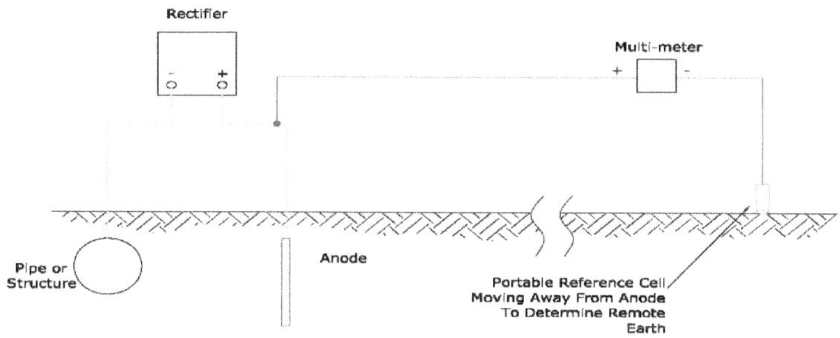

Figure 3-1: Remote earth distance measured using one reference cell.

As shown in Figure 3-1, the positive terminal of the multimeter is connected to the groundbed header cable and the negative terminal is connected to a reference cell. The potential is measured by moving the reference cell away from the groundbed.

The distance regarded as remote, for practical purposes, is that distance where the corresponding potential is no longer a significant change in the potential reads—5% or less. Figure 3-2 shows an actual graph obtained during the commissioning of a horizontal groundbed.

Freshour remote distance determination

Distance from last anode (ft)	Potential meas. (volt)
5	0.066
10	0.134
15	0.212
20	0.288
40	0.507
60	0.678
80	0.796
100	0.898
120	0.985
140	1.052
160	1.12
180	1.179
200	1.225
220	1.275
240	1.3
260	1.34
280	1.36
300	1.375
320	1.4
340	1.42
360	1.433
380	1.439
400	1.44
420	1.45
440	1.478
460	1.492
480	1.503
600	1.532

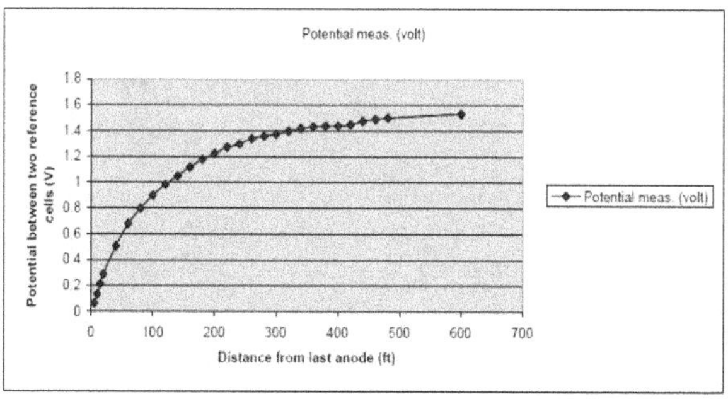

Percentage of potential change between 600 ft and 480 ft: 1.929474
Remote distance: 600 ft
Potential to remote earth at 332': 0.083 Volt
Percentage of potential change between remote earth (600 ft from the last anode) and minimum remote distance calculated during design calculations (332'): 5.42
The actual distance selected for groundbed was 460 ft,
So we can regard that the groundbed is remote from the pipeline.
Potential to remote earth 460' from last anode: 0.040 Volt
Percentage of potential change between remote earth (600 ft from the last anode) and actual distance between first anode and pipeline, 460 ft: 2.610966

Figure 3-2: Actual field measurements at an existing groundbed.

The table and graphic shown Figure 3-2 provide actual field values measured using the procedure shown on Figure 3-1 above. At the end of the table, note the small difference between the potential measured at 480 ft (150 m) and 600 ft (182 m). This is also shown on the graph to the right of the table. So, adding more space between the anode and the reference electrode does not increase the potential value, which indicates you have arrived at remote earth.

In this example, ξ (remote earth) was calculated at 332 ft (100 m). The percent of influence at 332 ft (100 m) is 5.42% for this example.

The calculated and field measured values show very good correlation.

When the remote earth is determined, the direction used to move away the reference cell should be free of buried metallic structures. Otherwise, the values obtained during this process can be affected by the influence of the current entering and/or leaving the buried metallic structure near the reference electrode.

3.2.2 Remote earth distance measured using two reference cells (Method 2)

The potential profile in the ground due to the groundbed can be measured by placing a reference cell connected to the positive terminal of a multimeter over the center of the groundbed, and using another reference cell connected to the negative terminal of the multimeter located at remote earth. This configuration is shown in Figure 3-3.

Utilizing this arrangement, the magnitude of the potential readings is positive. The electrode connected to the positive of the multimeter moves away from the groundbed toward the electrode located at remote earth (connected to the negative of the multimeter), obtaining the potential profile of the influence of the groundbed in its surroundings with respect to the distance from the positive electrode to the groundbed, y in this instance. Figure 3-4 represents the procedure to follow in the field.

The location of the reference electrode connected to the negative terminal was determined in the previous section, for the example using Figure 3-2. Remote earth measured at 600 feet (182 m) from the groundbed.

Placing the reference electrode that connects to the negative terminal of the multimeter at the remote earth location (600-ft for this example) as described above the second reference electrode is connected to the positive terminal of the multimeter and placed at the center of the groundbed. If we then start moving the reference electrode connected to the positive terminal of the multimeter towards the reference electrode placed at remote earth, we obtain a similar profile of the potential ***in the ground due*** to the groundbed as shown in Figure 3-2.

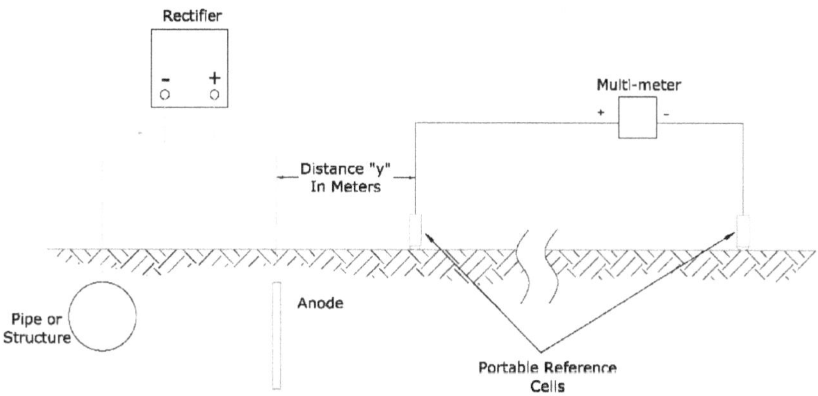

Figure 3-3: Potential profile measured using two reference cells.

3.2.3 The potential in the ground at the pipeline location due to the groundbed influence

The influence of an existing groundbed on the pipeline, can be determined using the following steps:

- Measure the distance from the pipeline to the groundbed. This is y in Figure 3-3 above.
- Place the reference electrode that is connected to the positive of the multimeter at the distance y from the groundbed, in the opposite direction to the pipe, as in Figure 3-3.
- Place the reference electrode that is connected to the negative of the multimeter at the remote earth distance, based on the calculation as determined in section 3.2.1, *i.e.*, the 600 ft value in Figure 3-2.
- The potential value measured is the potential that would have been calculated with any of the equations described below in Section 3.3, based on the corresponding groundbed configuration.
- If the value for y is ξ (remote distance as determined in Section 3.1), the potential measured with this procedure will be equal to 0.05 times, or 5%, the potential of the groundbed to remote earth.
- The potential value obtained as per Figure 3-3 is the voltage drop (IR) with respect to remote earth that the groundbed is adding to the protected pipeline potential at that distance from the groundbed, or the IR a foreign line may be experiencing in the proximity of the groundbed. See Figure 3-4 for the case of the calculated potential profile due to the groundbed and Figure 3-5 for an example of the pipe to soil potential with additional IR due to the presence of a groundbed. The pipeline in this example is not remote from the groundbed.

When the potential profile due to a groundbed is measured, and the reference cells are connected as in Figure 3-3, the potential values are positive if no external influences (stray current) are present.

It is essential that the potential difference between the two reference cells, determined before the field measurements, is noted so that corrections of the potential value measured between the two can be performed. Ideally, the two reference electrodes should be calibrated to a standard reference electrode and have a potential difference of 5mV.

The measurements as shown in Figure 3-3 are *not* potential gradients, but potential differences between the two reference electrodes. To obtain potential gradient values, you must divide the potential difference between two reference cells by the distance between them and consider the direction.

To be clear: the term "potential gradient" is often indiscriminately used in literature when what is actually being referred to are "potential difference" values. Potential gradients are vectors and are defined by the magnitude and the directions of the measured potential difference. The units of potential gradients are V/ (unit of length), like an electric field.

Since the potential profiles around the groundbed are symmetrical with respect to the groundbed axis, whatever values found in the direction opposite to the protected structure is the same in the direction of the protected structures (Figure 3-3). If you measure the potentials in the direction of the protected structure, the presence of the negative potentials generated in the ground due to the protected structure affects the reads taken to determine the groundbed potential profile and introduce measurement error. See Figure 3-6 with the simulation of the potential profile due to a gas well and the impressed groundbed that is protecting it as an example of what is explained in this paragraph.

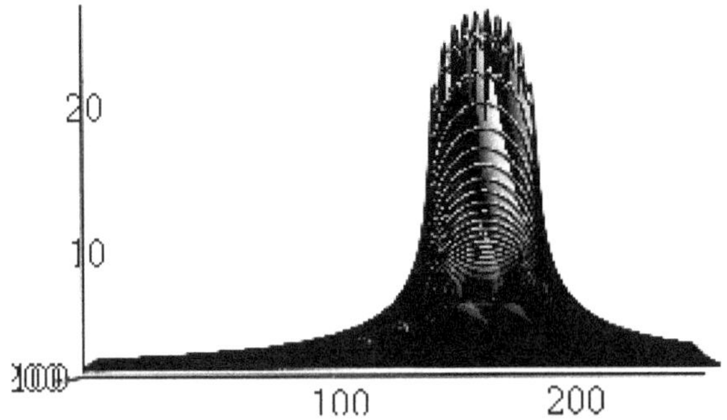

Figure 3-4: Potential profile calculated using an equation for horizontal groundbed.

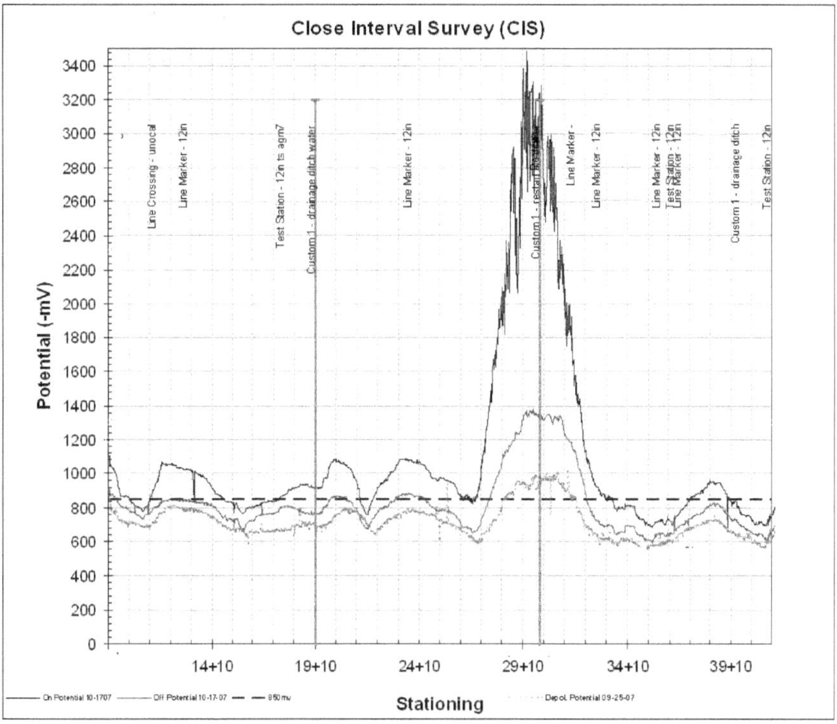

Figure 3-5: Pipe to soil CIS profile with the pipe under the influence of the groundbed (IR). Pipe is not remote from the groundbed.

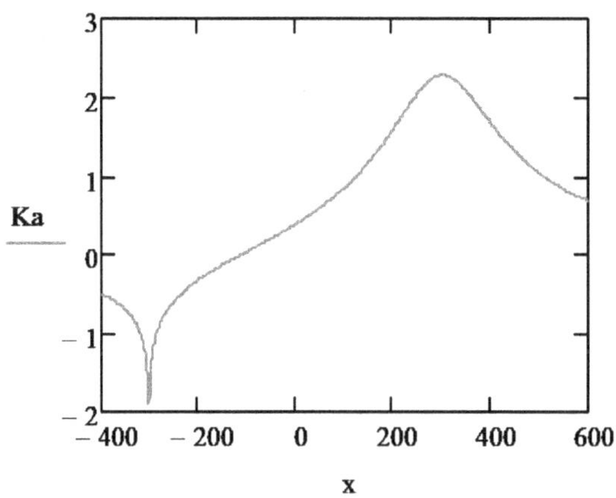

Figure 3-6: Combined potential profile calculated for a gas well (negative potential values) and its impressed current groundbed (positive values). All values were calculated to remote earth.

3.3 The potential on the ground due to vertical electrodes in various groundbed configurations

In these calculations, the x-axis is for the structure that is being protected. The groundbeds could be perpendicular to the pipeline and then will be on the y-axis, or they can be parallel to the pipeline at a certain distance from the pipe. Each of these scenarios is discussed below.

3.3.1 The potential on the surface of the ground due to vertical electrodes perpendicular to the pipeline

The following figure shows the configuration for this case.

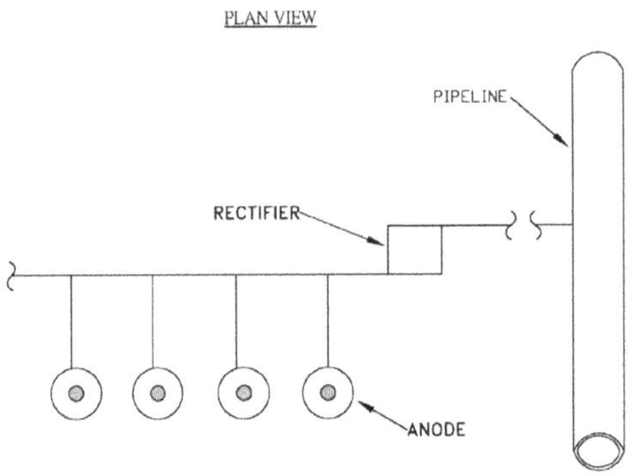

Figure 3-7: Plan view of vertical anodes installed perpendicular to the pipeline.

The equation for a *single vertical anode* potential was developed in Chapter 2 for various cases. First, we evaluate the instance when the anodes are at depth $t \neq 0$, as presented in Chapter 2, Section 2.2, Equations 2.3 and 2.4.

In general, the potential can be determined at any depth, where z represents the depth of the anode:

The potential equation, due to a single anode is

$$V(x, y, z) = \frac{I \cdot \rho}{4 \cdot \pi\, L} \cdot \left[\ln\left[\frac{z - t + \sqrt{x^2 + y^2 + (z - t)^2}}{z - t - L + \sqrt{x^2 + y^2 + (z - t - L)^2}} \right] + \ln\left[\frac{t + L + z + \sqrt{x^2 + y^2 + (t + L + z)^2}}{t + z + \sqrt{x^2 + y^2 + (t + z)^2}} \right] \right]$$

Equation 3.1

This is the same equation 2.3, in Chapter 2 with $r = \sqrt{x^2 + .y^2}$

Now, assume that the pipe is in the *x*-axis, at a depth of *z* and the anode is at the distance ξ, remote from *x*-axis in *y*-axis, perpendicular to the protected structure. Then Figure 3.8 shows the potential equation due to this anode is:

$$V(x,y,z) = \frac{Ides \cdot \rho}{4 \cdot \pi \cdot L} \cdot \left[\ln\left[\frac{z - t + \sqrt{(x)^2 + (y - \xi)^2 + (z - t)^2}}{z - t - L + \sqrt{(x)^2 + (y - \xi)^2 + (z - t - L)^2}} \right] + \ln\left[\frac{t + L + z + \sqrt{(x)^2 + (y - \xi)^2 + (t + L + z)^2}}{t + z + \sqrt{(x)^2 + (y - \xi)^2 + (t + z)^2}} \right] \right]$$

Equation 3.2

If we have a second anode, perpendicular to the pipe in the y-axis, at the distance *S* from the first anode, the distance from the pipe is ξ+*S*, and the potential equation due to the second anode is:

$$V2(x,y,z) = \frac{I \cdot \rho}{4 \cdot \pi\ L} \cdot \left[\ln\left[\frac{z - t + \sqrt{x^2 + (y - \xi)^2 + (z - t)^2}}{z - t - L + \sqrt{x^2 + (y - \xi)^2 + (z - t - L)^2}} \right] + \ln\left[\frac{t + L + z + \sqrt{x^2 + (y - \xi)^2 + (t + L + z)^2}}{t + z + \sqrt{x^2 + (y - \xi)^2 + (t + z)^2}} \right] \right]$$

Equation 3.3

The *y* coordinate of the second anode moves away from the pipe axis ξ+*S*.

In the case of just one anode, *I* enters that single anode only. When you have *N* number of anodes connected in parallel in the same groundbed, the total current *I* is shared between all the anodes.[5] Here, we are assuming that there is neither end effect nor attenuation in the current distribution in the group of anodes. This assumption is more accurate when two positive wires are connected between the rectifier junction box and the first and last anode of the groundbed, or when only one positive cable is connected to the midpoint on a header cable of a groundbed.

The total potential due to the two anodes is the sum of Equations 3.2 and 3.3, with *N*=2.

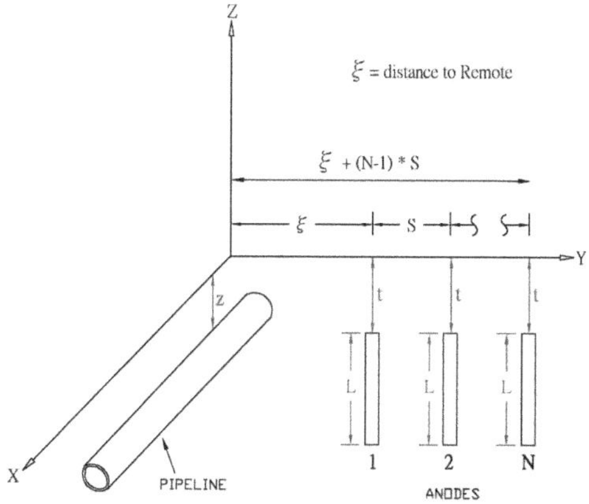

Figure 3-8: Vertical anodes perpendicular to the pipeline arrangement with *N* number of anodes.

If now we have a multiple number of anodes *N*, and with $o = \dfrac{I \cdot \rho}{4 \cdot \pi \; N \cdot L}$

The potential is:

$$V(x,y,z) = o \cdot \left[\sum_{n=0}^{N-1} \left[\left[\ln\left[\frac{z - t + \sqrt{x^2 + (y - \xi - n \cdot S)^2 + (z - t)^2}}{z - t - L + \sqrt{x^2 + (y - \xi - n \cdot S)^2 + (z - t - L)^2}} \right] + \ln\left[\frac{t + L + z + \sqrt{x^2 + (y - \xi - n \cdot S)^2 + (t + L + z)^2}}{t + z + \sqrt{x^2 + (y - \xi - n \cdot S)^2 + (t + z)^2}} \right] \right] \right] \right]$$

Equation 3.4

This equation considers the additive effect of all anodes.[6] The influence of all the anodes at a point in the soil is the addition of the influence of each anode.

When the potential is required in the surface of the ground, the equation takes the form:

$$V(x,y) = \sum_{n=0}^{Nh-1} \left[o \cdot \left[\ln\left[\frac{t + L + \sqrt{x^2 + (y - n \cdot S - \xi)^2 + (t + L)^2}}{t + \sqrt{x^2 + (y - n \cdot S - \xi)^2 + t^2}} \right] \right] \right]$$

Equation 3.4..

Calculating Equation 3.4.1 for several vertical anodes is shown in Figure 3-9. Assuming *z=0*, the potential profile provides the potential values on the surface of the earth. If the potential in the ground at the pipe depth is required, the pipeline's depth can be substituted for the value of *z* in Equation 3.4. The potential with *z=0* can be measured in the field to corroborate the simulation calculations.

In this example, the groundbed has ten anodes distributed in the *y*-axis. The remote distance ξ in the *y*-axis is utilized, thereby showing the groundbed far to the right, away from the *x*-axis by ξ amount.

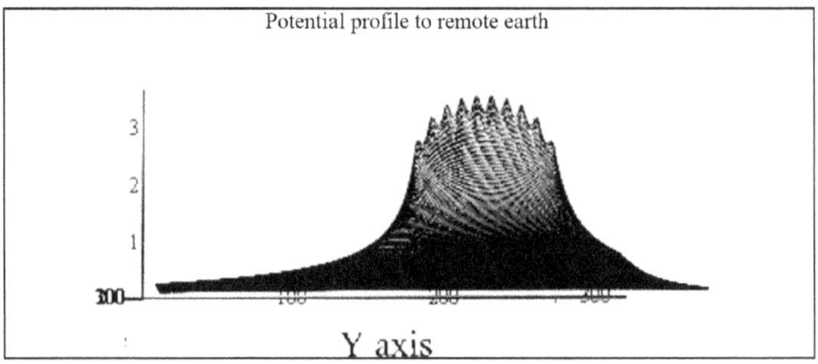

Ma

Figure 3-9: Potential profile for ten vertical anodes arranged perpendicular to the pipeline, which is in the x-axis entering the sheet of paper. The value of remote earth -ξ in this example is 170 ft.

3.3.2 The potential at the surface of the ground due to vertical electrodes parallel to the pipeline

Figure 3-10 shows a vertical parallel groundbed arrangement with respect to the pipeline.

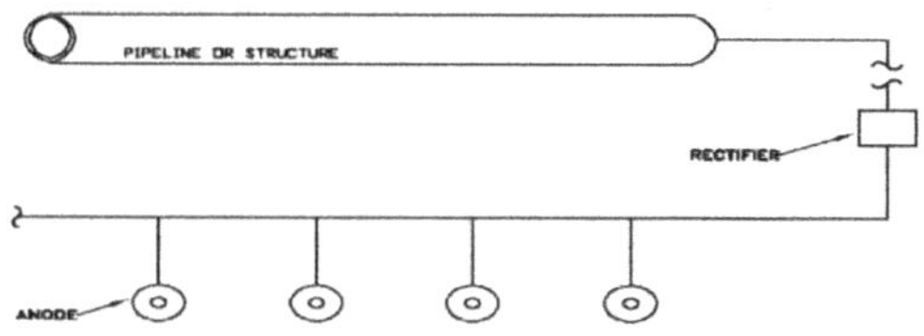

Figure 3-10: Plan view of vertical anodes installed parallel to the pipeline.

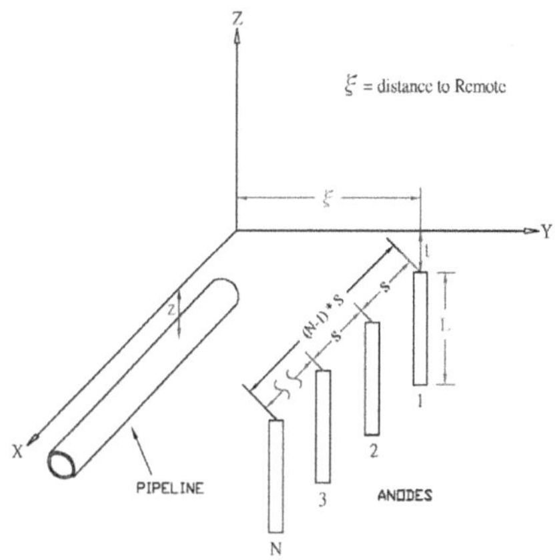

Figure 3-11: Elevation view of vertical anodes arranged parallel to the pipeline.

Equation 3.5 can be used to evaluate the groundbed influence:

$$V(x,y,z) = \rho \cdot \left[\sum_{n=0}^{N-1} \left[\ln\left[\frac{z - t + \sqrt{(x - n \cdot S)^2 + (y - \xi)^2 + (z - t)^2}}{z - t - L + \sqrt{(x - n \cdot S)^2 + (y - \xi)^2 + (z - t - L)^2}} \right] + \ln\left[\frac{t + L + z + \sqrt{(x - n \cdot S)^2 + (y - \xi)^2 + (t + L + z)^2}}{t + z + \sqrt{(x - n \cdot S)^2 + (y - \xi)^2 + (t + z)^2}} \right] \right] \right]$$

Equation 3.5

The variables are defined as before and input into Equation 3.5 results in the following graphic. (Figure 3-12 and Figure 3-13)

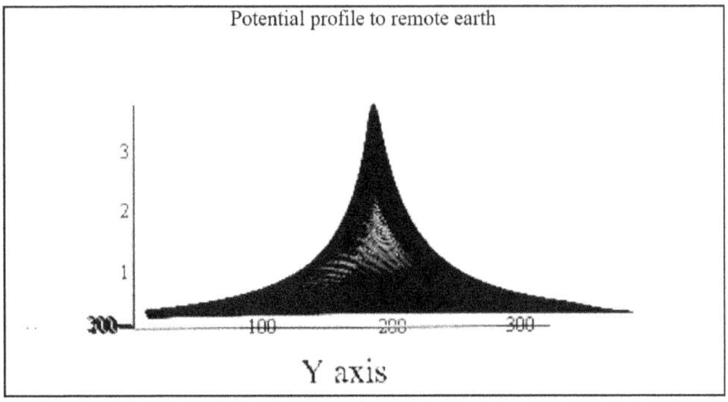

Ma

Figure 3-12: Potential profile for ten vertical anodes arranged parallel to the pipeline, which is on the *x*-axis entering the page. The value of ξ in this example is 170 ft.

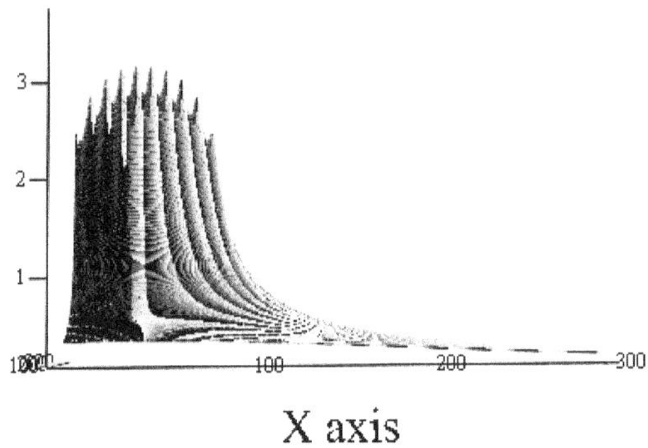

Figure 3-13: Potential profile for ten vertical anodes arranged parallel to the pipeline, which is on the *x*-axis. The *y*-axis is entering the page.

If the potential is required at the surface of the ground, the equation would take this form:

$$V(x,y) = 2\rho\cdot\left[\sum_{n=0}^{N-1}\left[\ln\left[\frac{t+L+\sqrt{(x-n\cdot S)^2+(y-\xi)^2+(t+L)^2}}{t+\sqrt{(x-n\cdot S)^2+(y-\xi)^2+t^2}}\right]\right]\right]$$

3.3.3 The potential in the ground due to vertical groundbed with the anodes in a straight line at a finite angle from the pipeline

For these calculations, the angles analyzed here and immediately below are measured with respect to the *x*-axis. In any design, the axes of coordinates can be chosen to facilitate the use of the equations, having the pipeline in the *x*-axis or parallel to this axis.

Now, we obtain a generalization for Equations 3.4 and 3.5 for anodes arranged vertically forming a straight line groundbed. This equation will combine Equations 3.4 and 3.5 to calculate the potential due to a vertical groundbed that forms any angle with the pipeline; particularly 90° (Equation 3.4) and 0° (Equation 3.5).

The new variables required for this calculation are:

x1,y1: the coordinates of the first anode of the groundbed; the closest one to the pipeline.

x2,y2: the coordinates of the centerline of the last anode in the groundbed; the furthest one from the pipeline.

Lg: the groundbed length. It depends on the number of anodes and distance between anodes as follows: Lg = N – 1)·S.

The hypotenuse coincides with the groundbed center line, and is defined as follows:

$$Hp1 = \sqrt{(x2 - x1)^2 + (y2 - y1)^2}$$

This value is equal to the length of the groundbed, *Lg*.

$$coseno = \frac{x2 - x1}{Hp1}$$

This is the cosine of the angle formed between the groundbed and the pipeline.

$$seno = \frac{y2 - y1}{Hp1}$$

This is the sine of the angle formed between the groundbed and the pipeline.

The other variables and constants have the same definition as stated previously.

The general equation is then:

$$\mathrm{Ui1}(x,y,z) = \sum_{n=0}^{N-1}\left[\sigma\cdot\ln\left[\frac{t+L+z+\sqrt{(x-n\cdot S\cdot coseno - x1)^2+(y-n\cdot S\cdot seno - y1)^2+(t+L+z)^2}}{z+t+\sqrt{(x-n\cdot S\cdot coseno - x1)^2+(y-n\cdot S\cdot seno - y1)^2+(z+t)^2}}\right]\right]$$

$$\mathrm{Ui2}(x,y,z) = \sum_{n=0}^{N-1}\left[\sigma\cdot\ln\left[\frac{z-t+\sqrt{(x-n\cdot S\cdot coseno - x1)^2+(y-n\cdot S\cdot seno - y1)^2+(z-t)^2}}{z-t-L+\sqrt{(x-n\cdot S\cdot coseno - x1)^2+(y-n\cdot S\cdot seno - y1)^2+(z-t-L)^2}}\right]\right]$$

$$\mathrm{Ui}(x,y,z) = \mathrm{Ui1}(x,y,z) + \mathrm{Ui2}(x,y,z)$$

Equation 3.6

The reader can calculate the extreme cases,for groundbed parallel to the pipeline, where *coseno=1*, and *seno=0*, and for groundbed perpendicular to the pipeline, where *coseno=0* and *seno=1* and compare the results with Equations 3.5 and 3.4, respectively.

3.4 The potential on the ground due to horizontal electrodes in various groundbed configurations

There are four different configurations for horizontal groundbeds.

3.4.1 Potential calculation due to horizontal groundbed perpendicular to the pipe, with the anodes parallel to the pipe:

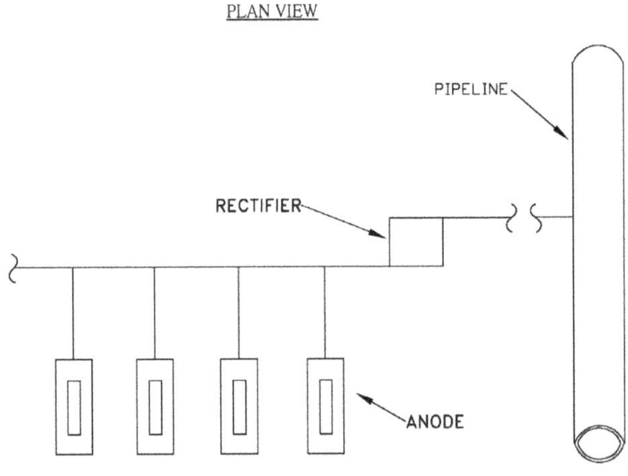

Figure 3.14: Plan view of horizontal anodes installed parallel to the pipeline with the groundbed perpendicular to the pipeline.

The equation takes the form, for the case of potential measured in the surface of the ground where $z=0$.

$$Vh(x,y) = \sum_{n=0}^{N-1} \left[2o \cdot \ln \left[\frac{x + \sqrt{x^2 + (y - \xi - n \cdot S)^2 + (t)^2}}{x - L + \sqrt{(x-L)^2 + (y - \xi - n \cdot S)^2 + (t)^2}} \right] \right] \qquad \text{Equation 3.7}$$

The determination of ξ, the remote earth distance, is performed using the same procedure in Section 3.1 above, but one needs to change the potential equation to Equation 3.7, inside the subprogram.

Evaluating for the general case of the equation for the potential at any depth, the equation has the form:

$$Vh(x,y,z) = o \cdot \sum_{n=0}^{N-1} \left[\ln \left[\frac{x + \sqrt{x^2 + (y - \xi - n \cdot S)^2 + (t+z)^2}}{x - L + \sqrt{(x-L)^2 + (y - \xi - n \cdot S)^2 + (z+t)^2}} \right] + \ln \left[\frac{x + \sqrt{x^2 + (y - \xi - n \cdot S)^2 + (z-t)^2}}{x - L + \sqrt{(x-L)^2 + (y - \xi - n \cdot S)^2 + (z-t)^2}} \right] \right] \qquad \text{Equation 3.8}$$

It is preferable to always have the option of equations that can be utilized to determine the potential in the surface of the ground, like Equation 3.7, to facilitate the possibility to corroborate the theory. Meanwhile equations that also consider a depth $t<0$ will provide the potential value for the examples to the pipe depth or to the depth of foreign structures, as in the case that interference is being evaluated to choose a better location for a groundbed.

3.4.2 The potential due to horizontal groundbed perpendicular to the pipe and the anodes perpendicular to the pipe

Figure 3-15 shows the groundbed configuration. This case is a generalization of Equation 2.11.1 from Chapter 2, where the anode and groundbed axis is the y-axis.

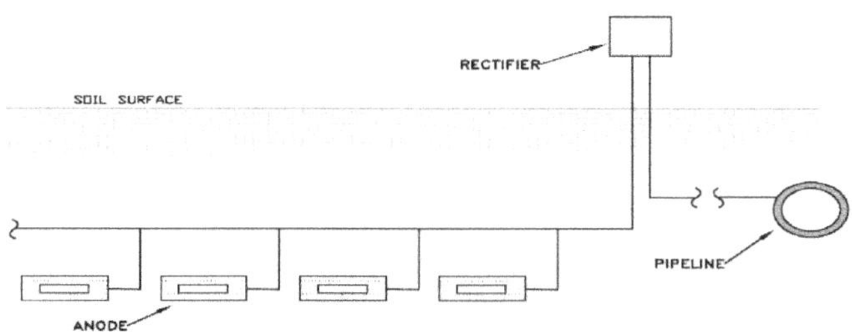

Figure 3-15: Individual horizontal anodes in a line perpendicular to a pipeline.

The general equation has the form:

$$Vh(x,y,z) = o\cdot\left[\sum_{n=0}^{N-1}\left[\ln\left[\frac{y-\xi-n\cdot S+\sqrt{x^2+(y-\xi-n\cdot S)^2+(t+z)^2}}{y-\xi-L-n\cdot S+\sqrt{x^2+(y-\xi-L-n\cdot S)^2+(t+z)^2}}\right]+\ln\left[\frac{y-\xi-n\cdot S+\sqrt{x^2+(y-\xi-n\cdot S)^2+(z-t)^2}}{y-\xi-L-n\cdot S+\sqrt{x^2+(y-\xi-L-n\cdot S)^2+(z-t)^2}}\right]\right]\right]$$

Equation 3.9

Where S=S1+L, S1 is the distance between the end of one anode and the start of the next anode and L is the length of the horizontal anode, and S1>0.

Moreover, the case of the potential at the surface of the ground:

$$Vh(x,y) = 2o\cdot\left[\sum_{n=0}^{N-1}\left[\ln\left[\frac{y-\xi-n\cdot S+\sqrt{x^2+(y-\xi-n\cdot S)^2+t^2}}{y-\xi-L-n\cdot S+\sqrt{x^2+(y-\xi-L-n\cdot S)^2+t^2}}\right]\right]\right]$$

Equation 3.10

The determination of ξ (remote earth distance) is again performed using the procedure as shown in previous cases but changing the potential equation to Equation 3.9 or Equation 3.10 in the subprogram.

3.4.3 The potential in the ground due to horizontal groundbed parallel to the pipe, the anodes perpendicular to the pipe

Figure 3-16 shows the groundbed configuration.

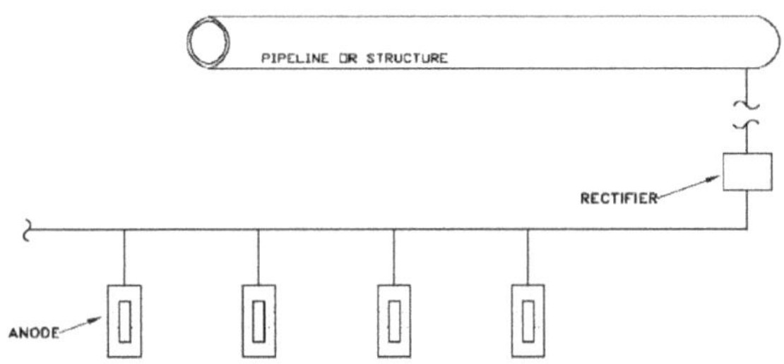

Figure 3-16: Horizontal groundbed parallel to the pipe and anodes perpendicular to the pipe.

The general equation has the form:

$$Vh(x,y,z) = o\cdot\left[\sum_{n=0}^{N-1}\left[\ln\left[\frac{y-\xi+\sqrt{(x-n\cdot S)^2+(y-\xi)^2+(t+z)^2}}{y-\xi-L+\sqrt{(x-n\cdot S)^2+(y-\xi-L)^2+(t+z)^2}}\right]+\ln\left[\frac{y-\xi+\sqrt{(x-n\cdot S)^2+(y-\xi)^2+(z-t)^2}}{y-L-\xi+\sqrt{(x-n\cdot S)^2+(y-\xi-L)^2+(z-t)^2}}\right]\right]\right]$$

Equation 3.11

For the case of the potential at the ground surface:

$$Vh(x,y) = 2o\cdot\sum_{n=0}^{N-1}\left[\ln\left[\frac{y-\xi+\sqrt{(x-n\cdot S)^2+(y-\xi)^2+t^2}}{y-\xi-L+\sqrt{(x-n\cdot S)^2+(y-\xi-L)^2+t^2}}\right]\right]$$

Equation 3.12

3.4.4 The potential in the ground due to horizontal groundbed, parallel to the pipe, the anodes parallel to the pipe

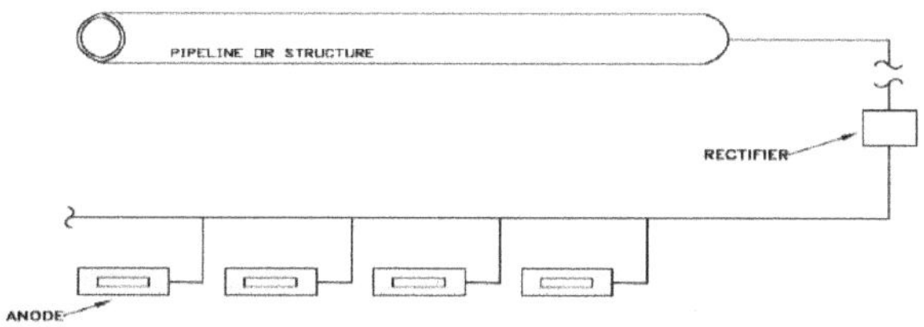

Figure 3-17: Horizontal anodes in a line parallel to the pipe.

The general potential equation has the form:

$$Vh(x,y,z) = \sum_{n=0}^{N-1}\left[o\cdot\left[\ln\left[\frac{x-n\cdot S+\sqrt{(x-n\cdot S)^2+(y-\xi)^2+(z-t)^2}}{x-L-n\cdot S+\sqrt{(x-L-n\cdot S)^2+(y-\xi)^2+(z-t)^2}}\right]+\ln\left[\frac{x-n\cdot S+\sqrt{(x-n\cdot S)^2+(y-\xi)^2+(z+t)^2}}{x-L-n\cdot S+\sqrt{(x-L-n\cdot S)^2+(y-\xi)^2+(z+t)^2}}\right]\right]\right]$$

Equation 3.13

Where S=S1+L, S1 is the distance between the end of one anode and the start of the next anode and L is the length of the horizontal anode, and S1>0.

For the case of the potential at the ground surface:

$$Vh(x,y) = \sum_{n=0}^{N-1}\left[o\cdot\ln\left[\frac{x-n\cdot S+\sqrt{(x-n\cdot S)^2+(y-\xi)^2+t^2}}{x-L-n\cdot S+\sqrt{(x-L-n\cdot S)^2+(y-\xi)^2+t^2}}\right]\right]$$

Equation 3.14

After the initially developed calculations from Chapters 1 and 2, the equations in Chapter 3 are geometric progressions to various practical applications.

3.4.5 The potential in the ground due to horizontal groundbed, where the anodes are in discrete coke beds, and the groundbed is at an angle from the pipeline

There are three cases to discuss here; when the horizontal anodes are parallel to the *x*-axis, when the horizontal anodes are parallel to the *y*-axis, and when the anodes are aligned in a straight line, at a certain angle from the *x*-axis.

3.4.5.1 The potential due to horizontal groundbed, where the anodes are in discrete coke beds, parallel to the pipeline and groundbed is at an angle from the pipeline

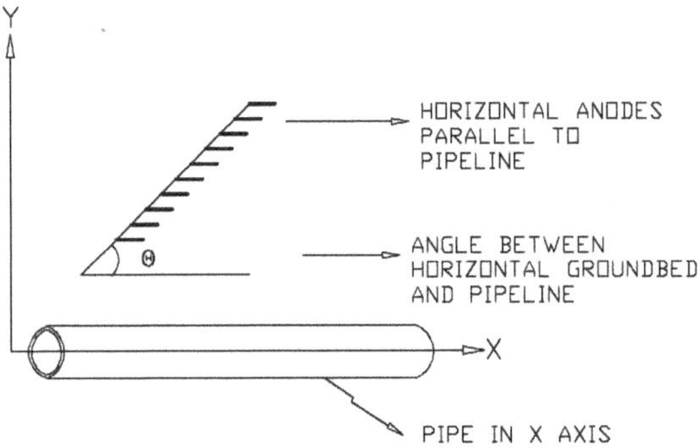

Figure 3-18: Horizontal groundbed, where the anodes are in discrete coke beds, parallel to the pipeline and groundbed is at an angle from the pipeline.

The cosine and sine of the angle between the groundbed and the pipeline were defined previously, as well as the procedure to obtain *x2* and *y2*.

The equation that describes the potential in the ground due to this configuration is:

$$Uh1(x,y,z) = \sum_{p=0}^{N-1}\left[o\cdot\ln\left[\frac{\sqrt{(x-p\cdot S\cdot coseno-x1)^2+(y-p\cdot S\cdot seno-y1)^2+(z-t)^2}+(x-p\cdot S\cdot coseno-x1)}{\sqrt{(x-p\cdot S\cdot coseno-x1-L)^2+(y-p\cdot S\cdot seno-y1)^2+(z-t)^2}+(x-L-p\cdot S\cdot coseno-x1)}\right]\right]$$

$$Uh2(x,y,z) = \sum_{p=0}^{N-1}\left[o\cdot\ln\left[\frac{\sqrt{(x-p\cdot S\cdot coseno-x1)^2+(y-p\cdot S\cdot seno-y1)^2+(z+t)^2}+(x-p\cdot S\cdot coseno-x1)}{\sqrt{(x-p\cdot S\cdot coseno-x1-L)^2+(y-p\cdot S\cdot seno-y1)^2+(z+t)^2}+(x-L-p\cdot S\cdot coseno-x1)}\right]\right]$$

Equation 3.15

$$Uh(x,y,z) = Uh1(x,y,z) + Uh2(x,y,z)$$

3.4.5.2. The potential in the ground due to horizontal groundbed, where the anodes are in discrete coke beds, perpendicular to the pipeline and groundbed is at an angle from the pipeline

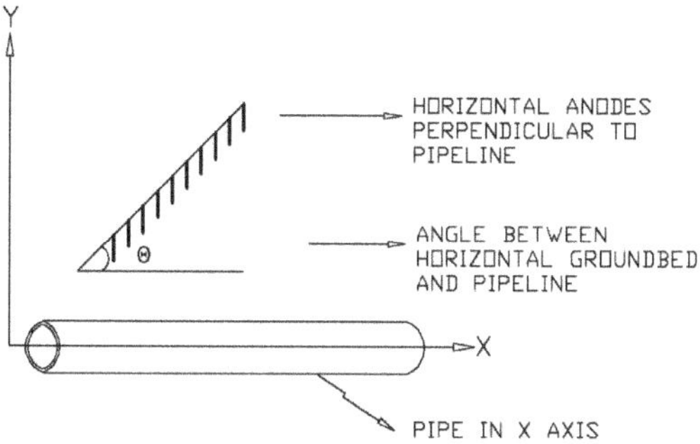

Figure 3-19: Horizontal groundbed, where the anodes are in discrete coke beds, perpendicular to the pipeline and groundbed is at an angle from the pipeline.

The equation that describes the potential due to this configuration is:

$$Uh1(x,y,z) = \sum_{p=0}^{N-1} \left[o \cdot \ln \left[\frac{\sqrt{(x - p \cdot S \cdot coseno - x1)^2 + (y - p \cdot S \cdot seno - y1)^2 + (z-t)^2} + (y - p \cdot S \cdot seno - y1)}{\sqrt{(x - p \cdot S \cdot coseno - x1)^2 + (y - L - p \cdot S \cdot seno - y1)^2 + (z-t)^2} + (y - L - p \cdot S \cdot seno - y1)} \right] \right]$$

$$Uh2(x,y,z) = \sum_{p=0}^{N-1} \left[o \cdot \ln \left[\frac{\sqrt{(x - p \cdot S \cdot coseno - x1)^2 + (y - p \cdot S \cdot seno - y1)^2 + (z+t)^2} + (y - p \cdot S \cdot seno - y1)}{\sqrt{(x - p \cdot S \cdot coseno - x1)^2 + (y - L - p \cdot S \cdot seno - y1)^2 + (z+t)^2} + (y - L - p \cdot S \cdot seno - y1)} \right] \right]$$

$$Uh(x,y,z) = Uh1(x,y,z) + Uh2(x,y,z)$$

Equation 3.16

3.4.5.3 The potential in the ground due to horizontal groundbed, where the anodes are in line, in discrete coke beds, groundbed is at an angle from the pipeline

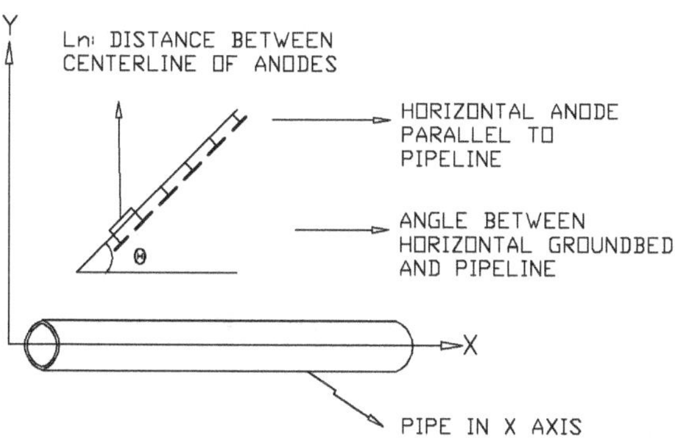

Figure 3-20: Horizontal groundbed, where the anodes are in discrete coke beds, in line and groundbed is at an angle from the pipeline

A new variable is defined for this case: Ln = S

This is the distance from center to center of each anode.

The integral equation was deducted from the integral equation used for the case of single horizontal anode at an angle from the pipeline, Chapter 2, Equation 2.15.

In this case, we must include the component: pLn

This component runs on the number of anodes from *0* to *N-1* and the distance between anodes Ln, as defined in the previous page.

The integral equations have the form:

$$Ug1(x,y,z) = o\cdot\left[\sum_{p=0}^{N-1}\int_{0}^{L}\left[\frac{1}{\sqrt{[x-x1-(u+p\cdot Ln)\cdot coseno]^{2}+[y-y1-(u+p\cdot Ln)\cdot seno]^{2}+(z-t)^{2}}}\right]du\right]$$

$$Ug2(x,y,z) = o\cdot\left[\sum_{p=0}^{N-1}\int_{0}^{L}\left[\frac{1}{\sqrt{[x-x1-(u+p\cdot Ln)\cdot coseno]^{2}+[y-y1-(u+p\cdot Ln)\cdot seno]^{2}+(z+t)^{2}}}\right]du\right]$$

$$Ug(x,y,z) = Ug1(x,y,z) + Ug2(x,y,z)$$

Equation 3.17

For the case of *z=0*, the only equation needed is:

$$Ug0(x,y,z) = 2\cdot o\cdot\left[\sum_{p=0}^{N-1}\int_{0}^{L}\left[\frac{1}{\sqrt{[x-x1-(u+p\cdot Ln)\cdot coseno]^{2}+[y-y1-(u+p\cdot Ln)\cdot seno]^{2}+(t)^{2}}}\right]du\right]$$

Equation 3.18

Like happened in this and previous chapters, we always must account for the electrode and its image; that is why we keep calculating components that have the elements *(z+t)* and *(z-t)*, respectively.

For those of you who do not have a computational program that allows the calculation of integral equations directly, the analytical equations follow.

Due to the length of the equations, we have identified the equation components:

The general components calculating the potentials at depth $z \neq 0$ are:

$$a11(x,y,z,p) = L^2 + (x - x1)^2 + (y - y1)^2 - 2 \cdot p \cdot Ln \cdot [(x - x1) \cdot coseno + (y - y1) \cdot seno]$$

$$a12(x,y,z,p) = p^2 \cdot Ln^2 + (z + t)^2 + 2L \cdot [p \cdot Ln - [(x - x1) \cdot coseno + (y - y1) \cdot seno]]$$

$$a1(x,y,z,p) = \sqrt{a11(x,y,z,p) + a12(x,y,z,p)}$$

$$b1(x,y,p) = [[p \cdot Ln - [(x - x1) \cdot coseno + (y - y1) \cdot seno]] + L]$$

$$c11(x,y,z,p) = [p \cdot Ln - [(x - x1) \cdot coseno + (y - y1) \cdot seno]]$$

$$c12(x,y,z,p) = \left[\sqrt{(x - x1)^2 + (y - y1)^2 - 2 \cdot p \cdot Ln \cdot [(x - x1) \cdot coseno + (y - y1) \cdot seno] + p^2 \cdot Ln^2 + (z + t)^2}\right]$$

$$c1(x,y,z,p) = c11(x,y,z,p) + c12(x,y,z,p)$$

$$a21(x,y,z,p) = L^2 + (x - x1)^2 + (y - y1)^2 - 2 \cdot p \cdot Ln \cdot [(x - x1) \cdot coseno + (y - y1) \cdot seno]$$

$$a22(x,y,z,p) = p^2 \cdot Ln^2 + (z - t)^2 + 2L \cdot [p \cdot Ln - [(x - x1) \cdot coseno + (y - y1) \cdot seno]]$$

$$a2(x,y,z,p) = \sqrt{a21(x,y,z,p) + a22(x,y,z,p)}$$

$$c21(x,y,z,p) = [p \cdot Ln - [(x - x1) \cdot coseno + (y - y1) \cdot seno]]$$

$$c22(x,y,z,p) = \left[\sqrt{(x - x1)^2 + (y - y1)^2 - 2 \cdot p \cdot Ln \cdot [(x - x1) \cdot coseno + (y - y1) \cdot seno] + p^2 \cdot Ln^2 + (z - t)^2}\right]$$

$$c2(x,y,z,p) = c21(x,y,z,p) + c22(x,y,z,p)$$

Please, notice that the component b1, a11 and c11 are the same for the electrode and its image.

$$Uae(x,y,z) = o \cdot \sum_{p=0}^{N-1} \left[\frac{(a1(x,y,z,p) + b1(x,y,p))}{c1(x,y,z,p)} + \frac{(a2(x,y,z,p) + b1(x,y,p))}{c2(x,y,z,p)}\right]$$

Equation 3.19

Equation 3.19 is the analytical solution of integral Equation 3.17.

Now, for the case of the potential calculated at the surface of the earth, where z=0

$$a01(x,y,p) = L^2 + (x - x1)^2 + (y - y1)^2 - 2 \cdot p \cdot Ln \cdot [(x - x1) \cdot coseno + (y - y1) \cdot seno]$$

$$a02(x,y,p) = p^2 \cdot Ln^2 + (t)^2 + 2L \cdot [p \cdot Ln - [(x - x1) \cdot coseno + (y - y1) \cdot seno]]$$

$$a0(x,y,p) = \sqrt{a01(x,y,p) + a02(x,y,p)}$$

$$c01(x,y,p) = [p \cdot Ln - [(x - x1) \cdot coseno + (y - y1) \cdot seno]]$$

$$c02(x,y,p) = \left[\sqrt{(x - x1)^2 + (y - y1)^2 - 2 \cdot p \cdot Ln \cdot [(x - x1) \cdot coseno + (y - y1) \cdot seno] + p^2 \cdot Ln^2 + (t)^2}\right]$$

$$c0(x,y,p) = c01(x,y,p) + c02(x,y,p)$$

The analytical equation has the form:

$$Uae0(x,y) = 2\cdot o\cdot \sum_{p=0}^{N-1} \left[\frac{(a0(x,y,p) + b1(x,y,p))}{c0(x,y,p)}\right] \qquad \text{Equation 3.20}$$

The Equations 3.17 and 3.18 can be used on calculation programs where integrals can be presented, like Mathcad or Matlab. The Equations 3.19 and 3.20 are the analytical solutions of Equation 3.17 and Equation 3.18 respectively and can also be used in Excel

REFERENCES

1. A.W. Peabody, R.L. Bianchetti, ed., *Peabody's Control of Pipeline Corrosion*, 2nd ed. (Houston, TX: NACE International, 2000), p. 29.
2. W. Von Baeckmann, W. Schwenk, W. Prinz, eds., *Handbook of Cathodic Corrosion Protection: Theory and Practice of Electrochemical Protection Processes*, 3rd ed. (Houston, TX: Gulf Publishing Company), p. 256.
3. J. Fitzgerald, J. Wagner, W Young, eds., *CP 4-Cathodic Protection Specialist Manual*, 1st ed. (Houston, TX: NACE International, 2000), p. 2:20.
4. J. Fitzgerald, J. Wagner, W Young, eds., *CP 4-Cathodic Protection Specialist Manual*, 1st ed. (Houston, TX: NACE International, 2000), p. 2:19.
5. E. D. Sunde, *Earth Conduction Effects in Transmission Systems*, 2nd ed. (New York, NY: Dover Publications, Inc., 1968), p. 78.
6. E. D. Sunde, *Earth Conduction Effects in Transmission Systems*, 2nd ed. (New York, NY: Dover Publications, Inc., 1968), p. 91.

Groundbed Resistance for Vertical and Horizontal Group of Anodes

To determine, the potential referenced to a point on the soil surface due to the influence of multiple anodes, the influence of each anode and their image over that point must be considered. This analysis was done without consideration of the interference between anodes.

When multiple anodes are in a groundbed configuration, they can be near each other such that the influence of the adjacent anodes should be considered in the groundbed's potential calculation. It means that the potential to remote earth of the groundbed, when the point of measurement coincides with the anodes surface needs to include a component due to this influence, which is referred to as ***mutual interference*** between anodes.

This influence increases the total potential of the groundbed and is dependent of the distances between anodes. In the case of a horizontal anode groundbed, the influence of the neighbor's image is also considered.

4.1 Resistance calculation for multiple anodes in vertical groundbed configuration.

Assumptions:

1. The groundbed potential is to remote earth.
2. The distance (spacing) between consecutive anodes, S, is the same.
3. Anodes are connected to the same positive terminal of the rectifier,
4. Anodes are in the same uniform soil, uniform soil resistivity.
5. The anodes are considered electrically in parallel.
6. When we use the term groundbed potential, it is the groundbed potential to remote earth.

7. The current being received by each anode is the total current injected from the rectifier divided by the total number of anodes. Attenuation of the current is not considered.
8. The average potential method is used, but instead of considering the length of the anodes to calculate the average, as is the case of individual anodes, the number of anodes is taken as the averaging factor. It will give the potential of one average anode.
9. The total groundbed potential to remote earth will be the potential of this average anode, mentioned in condition 8, divided by the number of anodes.
10. The sum of all anode resistances also includes the interference factor between anodes. All the anodes influence each other.
11. The radius of the anode is considered much less than the anode's length L and that the distance between anodes S: $a \ll L$ and $a \ll S$
12. The electrode charge is considered in the electrode's axis.

The first component in determining the total resistance of the groundbed is the resistance of an individual anode.

This term is obtained by the Equation 1.12 in Chapter 1.

$$Rp = \frac{1}{2 \cdot \pi} \cdot \frac{\rho}{L} \cdot \left(\ln\left(4 \cdot \frac{L}{a} \right) - 1 \right) \qquad \text{Equation 1.12}$$

The second term is due to the interference factor between anodes, it is known in some literature as the "crowding factor".

When anodes are placed very close to each other, the potential generated in the soil by each affects the others. This influence must be added to each anode's potential and to the total groundbed's potential.

When the potential of a single anode is calculated, without other influence, the charge is considered in the axis of the anode and the potential calculated at the surface of the anode. That is why the anode radius is used as the distance between the charges and the point where the potential is calculated. This is the method used in chapter one to obtain the Equations 1.9 and 1.12.

In the case of the influence of an anode due to the adjacent anodes, as in groundbeds, the distance from the anode to the charge will be the distance between the anode's axis and the axis of the other anodes. For this scenario we consider the conditions 3, 8, and 9, mentioned above. If the distance between anodes is much larger than the anode's radius, the distance between the anode and the other anodes in the groundbed will be multiple of S, because the charges are considered concentrated in the anode's axis.

When the interaction of the potential is considered between consecutive anodes, the distance to calculate the potential influence will be *S*. The total amount of terms will be *2(N-1)*, where N is the number of anodes.

When the interaction is considered between the anodes at distance *2S*, the number of terms is *2(N-2)*. The number 2 appears in the total number of terms based on condition 12 above; the anodes interaction is mutual.

When the interaction is considered between the anodes at distance *3S*, the number of terms is *2(N-3)*. And so on.

When the interaction is considered between the anodes at the end of the groundbed, anode *1* and anode *N*, at distance *S(N-1)*, the number of terms is *2*.

Because the distance between anodes is not much larger than the length of the anode, the equation used to calculate the resistance due to the influence between anodes is Equation 1.11, Chapter 1.

$$R = \frac{1}{2\cdot\pi}\cdot\frac{\rho}{L}\cdot\left[\ln\left[\frac{2L}{a}\cdot\left[\sqrt{1+\left(\frac{a}{2L}\right)^2}+1\right]\right]+\frac{a}{2\cdot L}-\sqrt{1+\left(\frac{a}{2L}\right)^2}\right] \quad \text{Equation 1.11}$$

The variable **a** represents the distance between the charges concentrated in the axis of the electrode and the surface of the electrode.

In this case to determine the influence between anodes, **a** will be substituted by *S*, *2S*, and so on, in function of what interference term is being calculated.

The influence can be expressed as follows:

$$R(S) = \frac{1}{2\cdot\pi}\cdot\frac{\rho}{L}\cdot\left[\ln\left[\frac{2L}{S}\cdot\left[\sqrt{1+\left(\frac{S}{2L}\right)^2}+1\right]\right]+\frac{S}{2\cdot L}-\sqrt{1+\left(\frac{S}{2L}\right)^2}\right] \quad \text{Equation 4.1}$$

The interference factor including the interaction between all the anodes that comply with this condition will be:

$$R(S)\cdot 2\cdot(N-1) \quad \text{Equation 4.2}$$

$$R(2S) = \frac{1}{2\cdot\pi}\cdot\frac{\rho}{L}\cdot\left[\ln\left[\frac{2L}{2S}\cdot\left[\sqrt{1+\left(\frac{2S}{2L}\right)^2}+1\right]\right]+\frac{2S}{2L}-\sqrt{1+\left(\frac{2S}{2L}\right)^2}\right] \quad \text{Equation 4.3}$$

The interference factor including the interaction between all the anodes that comply with this condition is:

$$R(2S)\cdot 2\cdot(N-2) \quad \text{Equation 4.4}$$

And so on.

Then, for the two anodes at the ends of the groundbed:

$$R(S(N-1)) = \frac{1}{2\cdot\pi}\cdot\frac{\rho}{L}\cdot\left[\ln\left[\frac{2L}{S(N-1)}\cdot\left[\sqrt{1+\left(\frac{S(N-1)}{2L}\right)^2}+1\right]\right]+\frac{S(N-1)}{2L}-\sqrt{1+\left(\frac{S(N-1)}{2L}\right)^2}\right]$$

Equation 4.5

The interference factor including the interaction between the anodes at the ends of the groundbed is:

$$R[S\cdot(N-1)]\cdot(2) \qquad \text{Equation 4.6}$$

The general term for n anodes is:

$$R(nS) = \frac{1}{2\cdot\pi}\cdot\frac{\rho}{L}\cdot\left[\ln\left[\frac{2L}{nS}\cdot\left[\sqrt{1+\left(\frac{nS}{2L}\right)^2}+1\right]\right]+\frac{nS}{2L}-\sqrt{1+\left(\frac{nS}{2L}\right)^2}\right]$$

Equation 4.7, where n runs from *1* to *N-1*

The total interference term, regarding all the anodes and the interaction between them is:

$$Rint = \frac{2}{N}\cdot\left[\sum_{n=1}^{N-1}[(N-n)\cdot R(nS)]\right] \qquad \text{Equation 4.8}$$

So, the total resistance of a groundbed with *N* anodes at spacing *S* between anodes, with uniform soil resistivity ρ, using the average potential method is:

$$Rg = \frac{1}{N}\cdot\left[\frac{1}{2\cdot\pi}\cdot\frac{\rho}{L}\cdot\left(\ln\left(4\cdot\frac{L}{a}\right)-1\right)+\frac{2}{N}\cdot\left[\sum_{n=1}^{N-1}[(N-n)\cdot R(nS)]\right]\right] \qquad \text{Equation 4.9}$$

The *N* at the beginning of the Equation 4.9 is due to the anodes are in parallel, so the total resistance of the groundbed is the resistance to remote earth of one anode, divided by the total number of anodes, because all the anodes are electrically in parallel.

The Equation 4.9 can also be arranged using Equation 2.7 for vertical anodes in Chapter 2 or with the equation for vertical anodes that appears in the Von Baeckmann book, Chapter 24, Equation 24-28.

4.2 Multiple parallel anodes in a horizontal configuration

We will not follow the Sunde's method for horizontal anodes, because he regards that the distance between the anodes and its image and the distance between consecutive parallel anodes is smaller than the length of the anodes.

Sunde's approach provides good correlation in the case of a single horizontal groundbed when the groundbed consists of several horizontal anodes in the same coke bed. In this case the length of the groundbed is always larger than the distance between the groundbed and its image.

Our approach follows that used by Dwight,[1] for the case of horizontal electrodes in parallel configuration, each in individual coke bed is shown in Figure 4-1

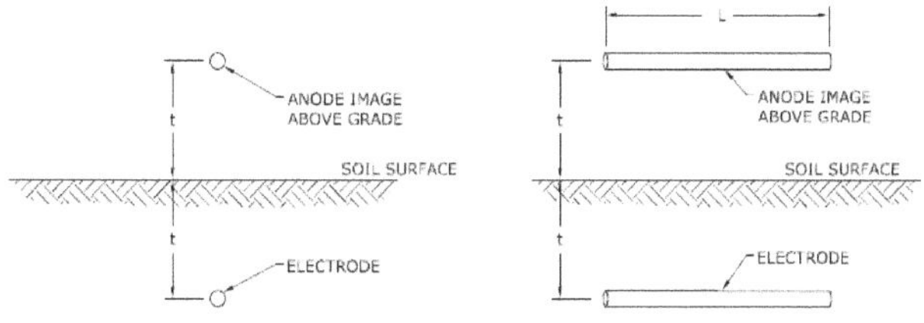

Figure 4-1: Individual horizontal anode and its image.

The equation obtained by a single horizontal electrode at depth *t*, with radius *a*, and length *L* is:

$$R0 = \frac{\rho}{2\pi \cdot L} \cdot \left[\ln\left(2 \cdot \frac{L}{a}\right) - 1 + \ln\left[\frac{L + \sqrt{(2 \cdot t)^2 + L^2}}{2 \cdot t}\right] + \frac{2 \cdot t}{L} - \frac{\sqrt{(2 \cdot t)^2 + (L)^2}}{L} \right]$$

From Equation 2.14, Chapter 2, Section 2.4.

If we write the equation as a function of the diameter instead of the radius **a**, Equation 2.14 is:

$$R0 = \frac{\rho}{2\pi \cdot L} \cdot \left[\ln\left(4 \cdot \frac{L}{d}\right) - 1 + \ln\left[\frac{L + \sqrt{(2 \cdot t)^2 + L^2}}{2 \cdot t}\right] + \frac{2 \cdot t}{L} - \frac{\sqrt{(2 \cdot t)^2 + (L)^2}}{L} \right]$$

To obtain an equation equivalent to that of Equation 4.9 above, for the case of multiple horizontal electrodes (anodes) in a groundbed, where the anodes are in individual coke bed, the following conditions should be considered.

1. The method of the average potential is applied; the number of anodes will be the quantity used for division to calculate the average value.
2. The anode's radius is much smaller than its length, the distance between anodes, and between anodes and images: $a \ll L$, $a \ll S$, $a \ll 2t$
3. All the charges in the anodes are at their respective axis.
4. The action of the anode's images, where these images are at a distance ***t*** over the surface of the earth is considered.

The spatial configuration of the anodes and their images is like two lines of anodes, in parallel with each other. One is below grade at depth ***t*** and the other is above grade at height ***t***. See Figure 4-2 below.

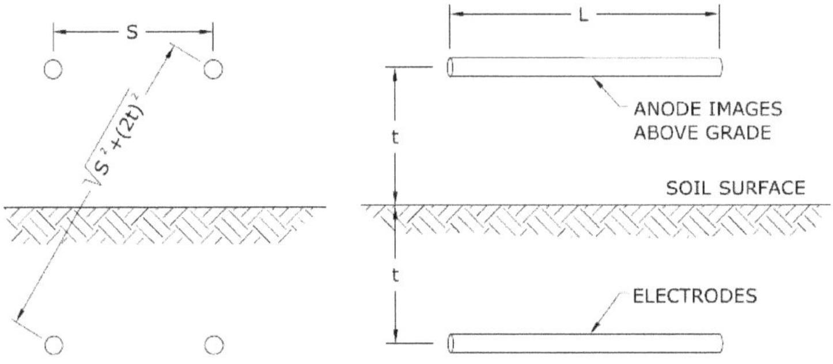

Figure 4-2: Two horizontal anodes and their images, connected in parallel.

The distance between the anode and the surface of the earth is *t*, the distance between anodes is *S*.

The distance between the anode and its image is *2t*.

The distance between the anode and the image of the adjacent anode is: $\sqrt{(2t)^2 + S^2}$. As shown in Figure 4-2.

As in the case of vertical anodes, the action between anodes is reciprocal. It means that whatever is occurring in one anode and its image, will happen in the other anode and its image.

Using Equation 2.13 derived in Chapter 2, if we use *S*, as the distance between one anode and its adjacent anode, the equation for the interference factor between two anodes can be expressed as follows:

$$\text{Rinta} = \frac{2\rho}{2\pi \cdot 2L} \cdot \left[\ln\left[\frac{L + \sqrt{(S)^2 + L^2}}{S} \right] + \frac{S}{L} - \frac{\sqrt{(S)^2 + (L)^2}}{L} \right] \quad \text{Equation 4.10}$$

Equation 4.10 includes multiplication by *2*, because the same interference will exist over the adjacent anode, and divided by two to obtain the average, based on the average potential method.

The interference factor between the anode and the image of the adjacent anode is:

$$\text{Rinti} = \frac{2\rho}{2\pi \cdot 2L} \cdot \left[\ln\left[\frac{L + \sqrt{\left[\sqrt{S^2 + (2t)^2}\right]^2 + L^2}}{\sqrt{S^2 + (2t)^2}} \right] + \frac{\sqrt{S^2 + (2t)^2}}{L} - \frac{\sqrt{\left[\sqrt{S^2 + (2t)^2}\right]^2 + L^2}}{L} \right]$$

Equation 4.11

Again, the Equation 4.11 is multiplied by 2, for the reciprocal effect between anodes, and divided by 2 to obtain the average.

Simplifying,

$$\text{Rinti} = \frac{2\rho}{2\pi \cdot 2L} \cdot \left[\ln\left[\frac{L + \sqrt{S^2 + (2t)^2 + L^2}}{\sqrt{S^2 + (2t)^2}} \right] + \frac{\sqrt{S^2 + (2t)^2}}{L} - \frac{\sqrt{S^2 + (2t)^2 + L^2}}{L} \right]$$

The total resistance of ***one average anode***, in a two parallel anode configuration, includes the sum of the following terms: Equations 2.14, 4.10 and 4.11. Because the anodes are electrically in parallel, it is divided by the total number of anodes, and the total groundbed resistance is obtained:

$$\text{Rh} = \frac{\text{R0} + \text{Rinta} + \text{Rinti}}{2} \qquad \text{Equation 4.12}$$

To prove this equation, we will use the example from the Dwight's paper,[2] in the case of two parallel buried horizontal electrodes, using the same values he used, but using Equation 4.12:

Anode's depth t=10 ft

Distance between anodes S=7 ft

Soil resistivity ρ=200,000 ohm ·cm

Anode diameter d=0.46 in

Anode length L=100 ft

First, using Dwight's method:

Dwight calculates the potential due to each rod, and each image, adds them and divide by the total charge of the four electrodes (two electrodes and two images):

$$\text{V0} = 2\left(\ln\left(4 \cdot \frac{L}{d}\right) - 1\right) \qquad \text{V0} = 16.5,$$

one rod of length L

$$\text{V1} = 2\left(\ln\left(\frac{L + \sqrt{S^2 + L^2}}{S}\right) + \frac{S}{L} - \frac{\sqrt{S^2 + L^2}}{L}\right)$$

$$\text{V1} = 4.84$$

at the rod, due to the other rod.

$$V2 = 2\cdot\left[\ln\left[\frac{L + \sqrt{(2t)^2 + L^2}}{2t}\right] + \frac{2t}{L} - \frac{\sqrt{(2t)^2 + (L)^2}}{L}\right]$$

$$V2 = 2.99$$

at the rod, due to its image.

$$V3 = 2\cdot\left[\ln\left[\frac{L + \sqrt{S^2 + (2t)^2 + L^2}}{\sqrt{S^2 + (2t)^2}}\right] + \frac{\sqrt{S^2 + (2t)^2}}{L} - \frac{\sqrt{S^2 + (2t)^2 + (L)^2}}{L}\right]$$

$$V3 = 2.89$$

at the rod, due to the image of the other rod.

The number 2 that appears in the above equations for potential implies that the action is reciprocal, the same action occurs in both directions. It was also applied in the case of vertical electrodes by Sunde.

Total charge:

$$Q = 4L$$

this is the charge of the two electrodes and two images.

$$Q = 12192 \text{ cm}$$

$$n = \frac{(V0 + V1 + V2 + V3)}{Q}$$

This is the capacitance of the system: potential divided by the total charge.

$$r = \frac{\rho}{2\cdot\pi}\cdot n$$

$$r = 71.078\text{ohm}$$

This is the resistance using the Equation 14 from Dwight's paper.

Now, using the Equation 4.12, with the input variables obtained above:

$$Rh = 71.078\text{ohm}$$

If we follow the APM, instead of changing the parameters in the Dwight equation, where we changed the anode's radius by the distance between anodes, calculating the influence of one anode in the adjacent anode using the potential equation, as it was explained in chapter 1, the procedure is as follows:

The horizontal anode potential influence in the surrounding is given by:

$$U(x,y,z) := o\cdot\left[\ln\left[\frac{x+\sqrt{(x)^2+(y)^2+(t+z)^2}}{x-L+\sqrt{(x-L)^2+(y)^2+(z+t)^2}}\right]+\ln\left[\frac{x+\sqrt{(x)^2+(y)^2+(z-t)^2}}{x-L+\sqrt{(x-L)^2+(y)^2+(z-t)^2}}\right]\right]$$

Equation 4.13

Where
$$o := \frac{Ia\cdot\rho}{4\cdot\pi\cdot L}$$

Equation 4.13 includes the influence on adjacent ground due to the anode and its image.

In this equation, x runs in the axis of the anode, y axis runs perpendicular to the anodes and z runs perpendicular to the earth surface, with negative values below grade.

The resistance interference factor between consecutive anodes will be obtained using the following integral and variables values:

$y := S \quad z := -t$ and x runs from x=0 to x=L

$$R12x := \frac{2o}{2\cdot Ia\cdot L}\cdot\int_0^L \ln\left[\frac{x+\sqrt{(x)^2+(y)^2+(t+z)^2}}{x-L+\sqrt{(x-L)^2+(y)^2+(z+t)^2}}\right]+\ln\left[\frac{x+\sqrt{(x)^2+(y)^2+(z-t)^2}}{x-L+\sqrt{(x-L)^2+(y)^2+(z-t)^2}}\right]dx$$

Equation 4.14

And with the values, as per the example above the value of the interference factor between horizontal anodes in a parallel configuration will be:

So, equation 4.14 is equivalent to equations 4.10 plus 4.11.

$$R12x = 40.381\cdot ohm$$

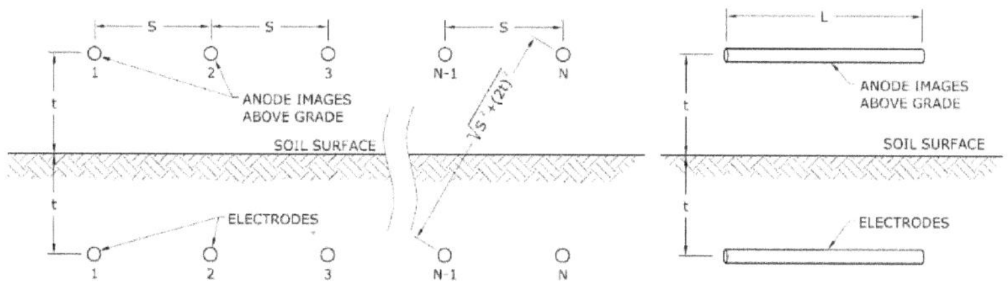

Figure 4-3: Generalization for the case of multiple horizontal anodes in a side-by-side configuration.

General term for the anode and its image, Equation 2.14:

$$\mathrm{Rinta} = \frac{\rho}{2\pi \cdot L} \cdot \left[\ln\left(4 \cdot \frac{L}{d}\right) - 1 + \ln\left[\frac{L + \sqrt{(2 \cdot t)^2 + L^2}}{2 \cdot t}\right] + \frac{2 \cdot t}{L} - \frac{\sqrt{(2 \cdot t)^2 + (L)^2}}{L} \right]$$

The general term between anodes, with distance between anodes *S*:

$$R1 = \frac{\rho}{2\pi} \cdot \frac{1}{L} \cdot \frac{2}{N} \cdot \left[\sum_{m=1}^{N-1} \left[(N-m) \cdot \left[\ln\left[\frac{L + \sqrt{(mS)^2 + L^2}}{mS}\right] + \frac{mS}{L} - \frac{\sqrt{(mS)^2 + (L)^2}}{L} \right] \right] \right]$$

Equation 4.15

General term between anode and images of other anodes:

$$R2 = \frac{\rho}{2\pi} \cdot \frac{1}{L} \cdot \frac{2}{N} \cdot \sum_{m=1}^{N-1} \left[(N-m) \cdot \ln\left[\left[\frac{L + \sqrt{(mS)^2 + (2t)^2 + L^2}}{\sqrt{(mS)^2 + (2t)^2}} \right] + \frac{\sqrt{(mS)^2 + (2t)^2}}{L} - \frac{\sqrt{(mS)^2 + (2t)^2 + (L)^2}}{L} \right] \right]$$

Equation 4.16

Remember the image of the anode has already been considered in Equation 2.14.

Equations 4.15 and 4.16 are divided by *N* (number of anodes) to account for the average potential for one anode and multiply by 2 to account for the reciprocal effect between anodes.

If we combine Equations 2.14, 4.15 and 4.16, the resistance of one average anode is obtained, and dividing by *N*, we obtain the resistance of the groundbed, considering that all the anodes are electrically in parallel.

The general equation for horizontal groundbeds is:

$$\mathrm{Rh} = \frac{R0 + R1 + R2}{N} \qquad \text{Equation 4.17}$$

Now, evaluating Equation 4.17 with the values from the example above, we have:

R0=101.8 ohm from Equation 2.14

R1=25.3 ohm from Equation 4.15

R2=15.1 ohm from Equation 4.16

Rht=71.078 ohm from Equation 4.17 with *N*=2

As expected, this value coincides with the value obtained using Equation 4.12 above.

4.3 Multiple horizontal electrodes (anodes) placed in a linear distribution, in discrete coke beds.

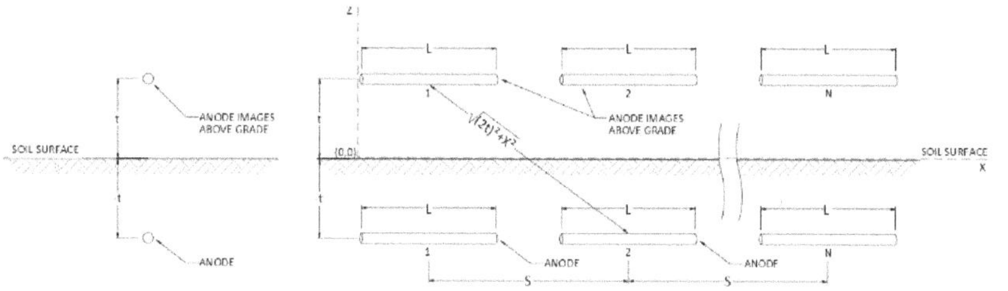

Figure 4-4: Number (*N*) of parallel horizontal anodes, arranged in a linear distribution (same trench).

In this case (Figure 4-4 above) the term related to interference between the anodes equivalent to Equations 4.15 and 4.16 should be obtained. We are applying the APM again, using the potential generic equation for horizontal anodes where the anodes are parallel to the x axis, like it is shown above to obtain equation 4.14:

The important factor here is that, using the same integral equation *we must change the integration limits and the values for y and z accordingly to adjust to the geometrical configuration*, and we will obtain a different resistance equation, now for the case of anodes in the same trench, but in discrete coke beds, in a linear configuration.

As per Figure 4-4, for the integration purpose, the origin of the coordinate system will be at the soil surface coinciding with the origin of the first anode for the x=0 coordinate.

The variables values will be as follows:

z is equal to -t, notice that here we only use z equal to -t due to the reciprocal effect, the anode influence on the other anode image is the same influence that the other anode image has on the first anode.

y is equal to 0, because all the anodes are in the same line.

Considering the influence of the first anode on the second anode, the integration limits will go from S to S+L. In the same way that S goes from the center of the first anode to the center of the second anode, *it goes from the beginning of the first anode to the beginning of the second anode, this facilitates the integration process.*

The integral form is as follows:

$$\text{Ra1LS} := \int_{S}^{L+S} \frac{2o}{2\cdot Ia\cdot L}\cdot\left[\ln\left[\frac{x+\sqrt{(x)^2+(y)^2+(t+z)^2}}{x-L+\sqrt{(x-L)^2+(y)^2+(z+t)^2}}\right]+\ln\left[\frac{x+\sqrt{(x)^2+(y)^2+(z-t)^2}}{x-L+\sqrt{(x-L)^2+(y)^2+(z-t)^2}}\right]\right] dx$$

Equation 4.18

Where o is obtained as follows:

$$o := \frac{Ia\cdot\rho}{4\cdot\pi\cdot L}$$

For the case that the reader cannot calculate the integral directly, the following analytical equations are equivalent:

$$\text{R1saS} := \frac{2o}{2\cdot Ia\cdot L}\cdot\left[(L+S)\cdot\ln\left(\frac{L+S}{S}\right)-(S-L)\cdot\ln\left(\frac{S}{S-L}\right)\right] \quad \text{Equation 4.19}$$

This is the component of interference related to the anode 1 over anode 2 and vice versa.

$$\text{R2saS1} := \frac{2o}{2\cdot Ia\cdot L}\cdot\left[(L+S)\cdot\ln\left[\frac{L+S+\sqrt{(L+S)^2+(2\cdot t)^2}}{S+\sqrt{(S)^2+(2\cdot t)^2}}\right]-(S-L)\cdot\ln\left[\frac{S+\sqrt{(S)^2+(2\cdot t)^2}}{S-L+\sqrt{(S-L)^2+(2\cdot t)^2}}\right]\right]$$

$$\text{R2saS2} := \frac{2o}{2\cdot Ia\cdot L}\cdot\left[2\cdot\sqrt{(S)^2+(2\cdot t)^2}-\sqrt{(S-L)^2+(2\cdot t)^2}-\sqrt{(L+S)^2+(2\cdot t)^2}\right]$$

$$\text{R2saS} := \text{R2saS1} + \text{R2saS2} \quad \text{Equation 4.20}$$

This is the component of interference related to the image of the first anode on the second anode and vice versa.

The total interference (crowded factor) component is then:

$$\text{R12saS} := \text{R1saS} + \text{R2saS} \quad \text{Equation 4.21}$$

The resistance equation for the case of a groundbed with only two anodes in a line configuration is obtained as follows:

$$\text{RhLsaS} := \frac{\text{R0} + \text{R12saS}}{\text{Ns}} \quad \text{Equation 4.22}$$

With Ns=2, Ns is the total number of anodes.

Where R0 is the resistance of one anode and its image, equation 2.14.

Now if we extrapolate for the case of many anodes, the integral form of interaction between anodes and images will be:

$$\mathrm{Raa1LS} := \sum_{m=1}^{Ns-1} \int_{m\cdot S}^{L+m\cdot S} \frac{2\rho\cdot(Ns-m)}{Ns\cdot Ia\cdot L}\cdot\left[\ln\left[\frac{x+\sqrt{(x)^2+(y)^2+(t+z)^2}}{x-L+\sqrt{(x-L)^2+(y)^2+(z+t)^2}}\right]+\ln\left[\frac{x+\sqrt{(x)^2+(y)^2+(z-t)^2}}{x-L+\sqrt{(x-L)^2+(y)^2+(z-t)^2}}\right]\right]dx$$

Equation 4.23, where m runs on from 1 to Ns-1.

$$\mathrm{RhL} := \frac{\mathrm{R0} + \mathrm{Raa1LS}}{Ns}$$

Equation 4.24 is the total resistance of all anodes in parallel.

In the analytical form:

$$\mathrm{R1saa} := \frac{\rho}{4\pi\cdot L}\cdot\frac{2}{Ns}\cdot\left[\sum_{m=1}^{Ns-1}\left[\frac{(Ns-m)}{L}\cdot\left[(L+m\cdot S)\cdot\ln\left(\frac{L+m\cdot S}{m\cdot S}\right)-(m\cdot S-L)\cdot\ln\left(\frac{m\cdot S}{m\cdot S-L}\right)\right]\right]\right]$$

Equation 4.25 this is the component of anode-anode interaction.

$$\mathrm{R2saa1} := \frac{\rho}{4\pi\cdot L}\cdot\frac{2}{Ns}\cdot\sum_{m=1}^{Ns-1}\left[\frac{(Ns-m)}{L}\cdot\left[(L+m\cdot S)\cdot\ln\left[\frac{L+m\cdot S+\sqrt{(L+m\cdot S)^2+(2\cdot t)^2}}{m\cdot S+\sqrt{(m\cdot S)^2+(2\cdot t)^2}}\right]-(m\cdot S-L)\cdot\ln\left[\frac{m\cdot S+\sqrt{(m\cdot S)^2+(2\cdot t)^2}}{m\cdot S-L+\sqrt{(m\cdot S-L)^2+(2\cdot t)^2}}\right]\right]\right]$$

$$\mathrm{R2saa2} := \frac{\rho}{4\pi\cdot L}\cdot\frac{2}{Ns}\cdot\sum_{m=1}^{Ns-1}\left[\frac{(Ns-m)}{L}\cdot\left[2\cdot\sqrt{(m\cdot S)^2+(2\cdot t)^2}-\sqrt{(m\cdot S-L)^2+(2\cdot t)^2}-\sqrt{(L+m\cdot S)^2+(2\cdot t)^2}\right]\right]$$

$$\mathrm{R2saa} := \mathrm{R2saa1} + \mathrm{R2saa2}$$

Equation 4.26, this is the component due to the image of one anode with the other anodes interaction.

The resistance of a groundbed with *Ns* anodes linearly distributed in discrete coke beds will be:

$$\mathrm{Rhhsaa} := \frac{\mathrm{R0} + \mathrm{R1saa} + \mathrm{R2saa}}{Ns} \qquad \text{Equation 4.27}$$

This last approach (anodes in a linear configuration in the same trench) is applicable for example when horizontal anodes need to be installed in trenches that are separated for driveways, where a single horizontal bed with multiple anodes cannot be installed.

It is important to mention that the distance between anodes S, in the case of anodes in the same trench is measured from the center of one anode to the center of the next anodes. It depends on the length of each anode and the distance between the end of one anode and the beginning of the next anode. For clarity we can call that distance S1.

Then

$$S = \frac{L}{2} + S1 + \frac{L}{2} = L + S1$$

REFERENCES

1. H.B. Dwight, "Calculation of Resistances to Ground," in Classic Papers and Reviews on Anode Resistance Fundamentals and Applications, ed. R.L. Benedict, (Houston, TX: NACE, 186), p. 3.
2. H.B. Dwight, "Calculation of Resistances to Ground," in Classic Papers and Reviews on Anode Resistance Fundamentals and Applications, ed. R.L. Benedict, (Houston, TX: NACE, 186), p. 5.

Resistance and Potential Equations for Groundbeds with Asymmetric Configurations

In Chapters 3 and 4 equations were developed to determine the potential in the ground due to a group of anodes in both vertical and horizontal configurations where the anodes have the same length, distance between anodes, and depth of the anodes.

In the real world there are instances where, due to the inhomogeneity of the soil or the presence of buried foreign structures, or restriction on the easement, the groundbeds cannot be installed in a uniform arrangement such that, the installation of a non-uniform distributed anodes groundbed may be necessary.

All these cases require different approaches to the traditional horizontal, vertical, or deep wells. Some of these options are presented in this chapter.

5.1 Resistance and potential calculation when the depth of the top, the length and spacing between the vertical anodes are not the same.

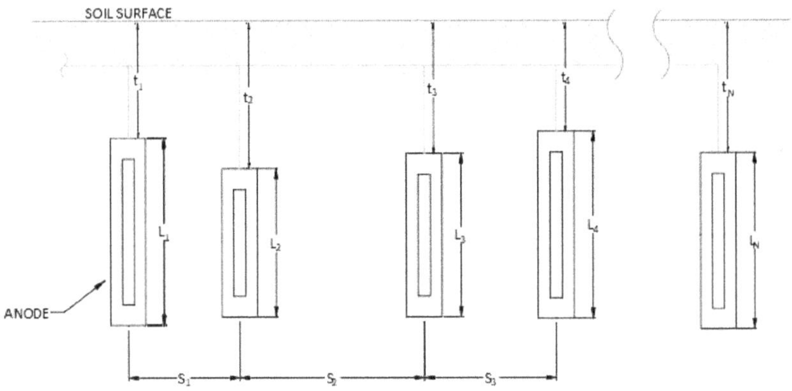

Figure 5-1: Example of vertical groundbed with anodes with different depth, length and spacing between anodes.

The main hypothesis used to obtain the potential and resistance equations have been presented in the previous chapters, here we will briefly discuss them again:

- The current is assumed to be uniformly distributed in the electrode length.
- The current is concentrated in the electrode's axis.
- The origin of the coordinate system is located at the soil surface with the z-axis coinciding with the first electrode axis.
- The electrodes length L_i is much larger than their radius $\boldsymbol{a}$.
- The integration variable is z that runs in the axis of the electrode, from $-L_i$-t_i, the bottom end of the electrode to $-t_i$, the top end of the electrode of length L_i. Where i varies from 1 to N, being N the number of electrodes (anodes in the case of cathodic protection) in the groundbed.
- The Average Potential Method (APM) is used to obtain the electrode's resistance to remote earth: the influence of the electrodes on each other and total groundbed resistance.
- The buried electrode and its image are electrically in parallel.

Two potential equations should be calculated to obtain the components of the total groundbed resistance: the potential equation corresponding to each anode's charge and the potential equation that considers the resistance component due to interference between parallel electrodes.

It is important to mention that the procedures that are following in this chapter even when they are applied to vertical anode systems, they also apply to horizontal anode systems.

Note: ***The APM regards that the total potential of an electrode to remote earth is the sum of the potential calculated at each point on the electrode surface divided by the length of the electrode in function of the current entering that electrode. In the case of several anodes in parallel, the APM regards the influence on an electrode due to the other electrodes and the division is by the total number of electrodes***[1]

5.1.1 Resistance component due to the anode's charge.

The equation for the resistance component due to the anode's charge is expressed regarding the APM with the following integral and iterative calculation [2].

$$Ga_j = \frac{\rho}{4\cdot\pi}\cdot\frac{1}{L_j}\cdot\frac{1}{L_j}\cdot\left[\int_{-(L_j+t_j)}^{-t_j}\left[\ln\left[\frac{L_j+t_j+z+\sqrt{(a)^2+(z+L_j+t_j)^2}}{z+t_j+\sqrt{(a)^2+(z+t_j)^2}}\right]+\ln\left[\frac{z-t_j+\sqrt{(a)^2+(z-t_j)^2}}{z-t_j-L_j+\sqrt{(a)^2+(z-t_j-L_j)^2}}\right]\right]dz\right]$$

Equation 5.1

If we use an iterative tool, for example, Mathcad®, it will look like:

$$Ga_j = \left[\begin{array}{l} \text{for } j \in 1..N \\ \quad Ga_j \leftarrow \frac{\rho}{4\cdot\pi}\cdot\frac{1}{L_j}\cdot\frac{1}{L_j}\cdot\left[\int_{-(L_j+t_j)}^{-t_j}\left[\ln\left[\frac{L_j+t_j+z+\sqrt{(a)^2+(z+L_j+t_j)^2}}{z+t_j+\sqrt{(a)^2+(z+t_j)^2}}\right]+\ln\left[\frac{z-t_j+\sqrt{(a)^2+(z-t_j)^2}}{z-t_j-L_j+\sqrt{(a)^2+(z-t_j-L_j)^2}}\right]\right]dz\right] \\ Ga_j \end{array}\right]$$

Equation 5.2

The variables are defined as follow:

Ga: electrode *j* contribution to the groundbed resistance in ohm
ρ: soil resistivity, in ohm-meter
L_j: length of the anode *j*, in meter
t_j: depth of the top of the anode *j* in meter
a: anode's radius in meter
N: is the total amount of anodes
z: integration variable that varies in the electrode axis from the bottom $(-t_j - L_j)$ of the electrode to the top of the electrode $(-t_j)$

The anode length *(L)* and anode depth *(t)* are provided using the following vectors, respectively:

Here, we have shown the vectors for the case of seven electrodes, as an illustrative example.

$$L = \begin{pmatrix} L1 \\ L2 \\ L3 \\ L4 \\ L5 \\ L6 \\ L7 \end{pmatrix} \cdot m \qquad t = \begin{pmatrix} t1 \\ t2 \\ t3 \\ t4 \\ t5 \\ t6 \\ t7 \end{pmatrix} \cdot m$$

To facilitate the calculations, the number of anodes is automatically taken as:

$$N = \text{rows}(L)$$

5.1.2 Resistance component due to interference between anodes.

The components to determine the crowding or interference factor between anodes are formed through a matrix approach.

To define the matrix we assume that the electrodes identified with the iterative term i, are the ones creating the interference on the electrodes with the iterative term *j*. Where column vector *S* represents the different distances (spacing) between the vertical electrodes.

There are three sets of values that conform the matrix of interference, and they are described as follow:

1. The components of the case when *i=j*, means we see the influence of anode *j+1* on the anode *j*. These are the diagonal components in the matrix of interference and can be expressed as follows:

$$Gat_{j,i} = \frac{\rho}{4 \cdot \pi} \cdot \frac{1}{L_i} \cdot \frac{1}{L_j} \cdot \int_{-(L_j+t_j)}^{-t_j} \left[\ln\left[\frac{L_i + t_i + z + \sqrt{(S_i)^2 + (z + L_i + t_i)^2}}{z + t_i + \sqrt{(S_i)^2 + (z + t_i)^2}} \right] + \ln\left[\frac{z - t_i + \sqrt{(S_i)^2 + (z - t_i)^2}}{z - t_i - L_i + \sqrt{(S_i)^2 + (z - t_i - L_i)^2}} \right] \right] dz$$

Equation 5.3

2. The components of the case when *j>i*. The anodes creating interference are *j-1* over *j*. These are the matrix elements to the left of the diagonal and are expressed as follows:

$$Gat_{j,i} = \frac{\rho}{4 \cdot \pi} \cdot \frac{1}{L_i} \cdot \frac{1}{L_j} \cdot \int_{-(L_j+t_j)}^{-t_j} \left[\ln\left[\frac{L_i + t_i + z + \sqrt{\left(\sum_{n=i}^{j-1} S_n\right)^2 + (L_i + t_i + z)^2}}{z + t_i + \sqrt{\left(\sum_{n=i}^{j-1} S_n\right)^2 + (z + t_i)^2}} \right] + \ln\left[\frac{z - t_i + \sqrt{\left(\sum_{n=i}^{j-1} S_n\right)^2 + (z - t_i)^2}}{z - t_i - L_i + \sqrt{\left(\sum_{n=i}^{j-1} S_n\right)^2 + (z - t_i - L_i)^2}} \right] \right] dz$$

Equation 5.4

3. The components of the case when $j<i$. The anodes creating the interference are $j+2$, $j+3$, to N, over the anode j. These are the matrix components right of the diagonal.

$$Gat_{j,i} := \frac{\rho}{4\cdot\pi}\cdot\frac{1}{L_i}\cdot\frac{1}{L_j}\cdot\int_{-(L_j+t_j)}^{-t_j}\left[\ln\left[\frac{L_i+t_i+z+\sqrt{\left(\sum_{n=j}^{i}S_n\right)^2+(L_i+t_i+z)^2}}{z+t_i+\sqrt{\left(\sum_{n=j}^{i}S_n\right)^2+(z+t_i)^2}}\right]+\ln\left[\frac{z-t_i+\sqrt{\left(\sum_{n=j}^{i}S_n\right)^2+(z-t_i)^2}}{z-t_i-L_i+\sqrt{\left(\left(\sum_{n=j}^{i}S_n\right)\right)^2+(z-t_i-L_i)^2}}\right]\right]dz$$

Equation 5.5

Here, we have shown the vectors for the case of seven electrodes, as an illustrative example.

$$L=\begin{pmatrix}L1\\L2\\L3\\L4\\L5\\L6\\L7\end{pmatrix}\cdot m \qquad t=\begin{pmatrix}t1\\t2\\t3\\t4\\t5\\t6\\t7\end{pmatrix}\cdot m \qquad S=\begin{pmatrix}S1\\S2\\S3\\S4\\S5\\S6\end{pmatrix}\cdot m$$

Now, combining the three equations in the matrix arrangement, using Mathcad®, the interference component of the groundbed resistance can be expressed as follows:

$$Gat_{j,i} := \left|\begin{array}{l}\text{for } j\in 1..N\\ \quad\text{for } i\in 1..N-1\\ \quad\quad Gat_{j,i}\leftarrow\frac{\rho}{4\cdot\pi}\cdot\frac{1}{L_i}\cdot\frac{1}{L_j}\cdot\int_{-(L_j+t_j)}^{-t_j}\left[\ln\left[\frac{L_i+t_i+z+\sqrt{(S_i)^2+(z+L_i+t_i)^2}}{z+t_i+\sqrt{(S_i)^2+(z+t_i)^2}}\right]+\ln\left[\frac{z-t_i+\sqrt{(S_i)^2+(z-t_i)^2}}{z-t_i-L_i+\sqrt{(S_i)^2+(z-t_i-L_i)^2}}\right]\right]dz \ \text{ if } j=i\\ \quad\quad Gat_{j,i}\leftarrow\frac{\rho}{4\cdot\pi}\cdot\frac{1}{L_i}\cdot\frac{1}{L_j}\cdot\int_{-(L_j+t_j)}^{-t_j}\left[\ln\left[\frac{L_i+t_i+z+\sqrt{\left(\sum_{n=i}^{j-1}S_n\right)^2+(L_i+t_i+z)^2}}{z+t_i+\sqrt{\left(\sum_{n=i}^{j-1}S_n\right)^2+(z+t_i)^2}}\right]+\ln\left[\frac{z-t_i+\sqrt{\left(\sum_{n=i}^{j-1}S_n\right)^2+(z-t_i)^2}}{z-t_i-L_i+\sqrt{\left(\sum_{n=i}^{j-1}S_n\right)^2+(z-t_i-L_i)^2}}\right]\right]dz \ \text{ if } j>i\\ \quad\quad Gat_{j,i}\leftarrow\frac{\rho}{4\cdot\pi}\cdot\frac{1}{L_i}\cdot\frac{1}{L_j}\cdot\int_{-(L_j+t_j)}^{-t_j}\left[\ln\left[\frac{L_i+t_i+z+\sqrt{\left(\sum_{n=j}^{i}S_n\right)^2+(L_i+t_i+z)^2}}{z+t_i+\sqrt{\left(\sum_{n=j}^{i}S_n\right)^2+(z+t_i)^2}}\right]+\ln\left[\frac{z-t_i+\sqrt{\left(\sum_{n=j}^{i}S_n\right)^2+(z-t_i)^2}}{z-t_i-L_i+\sqrt{\left(\left(\sum_{n=j}^{i}S_n\right)\right)^2+(z-t_i-L_i)^2}}\right]\right]dz \ \text{ if } j<i\\ Gat_{j,i}\end{array}\right.$$

Equation 5.6

Now we will calculate this matrix utilizing a practical example:

Soil resistivity ρ: 30 ohm-meter
Number of vertical anodes N: 7
Anode's diameter d: 0.203 m

$$L = \begin{pmatrix} 7 \\ 8 \\ 6 \\ 4 \\ 3 \\ 9 \\ 7 \end{pmatrix} m \quad t = \begin{pmatrix} 1 \\ 3 \\ 2 \\ 4 \\ 5 \\ 2 \\ 3 \end{pmatrix} m \quad S = \begin{pmatrix} 3 \\ 4 \\ 6 \\ 2 \\ 5 \\ 7 \end{pmatrix} \cdot m$$

The matrix of interference has the following values:

$$Gat = \begin{pmatrix} 0.90414456 & 0.48707406 & 0.32819616 & 0.28694085 & 0.22253496 & 0.16913616 \\ 0.78125554 & 0.66472011 & 0.38125628 & 0.32986209 & 0.24901422 & 0.18493694 \\ 0.52735873 & 0.68187894 & 0.58183519 & 0.46499191 & 0.31836843 & 0.2208841 \\ 0.32285059 & 0.37771108 & 0.56821876 & 1.18525805 & 0.5012805 & 0.2947483 \\ 0.28459224 & 0.32781666 & 0.45838308 & 1.19289591 & 0.63695878 & 0.33001122 \\ 0.22152822 & 0.24777031 & 0.31594907 & 0.48871806 & 0.60315809 & 0.47637891 \\ 0.16936176 & 0.18520019 & 0.22132342 & 0.29566707 & 0.33132919 & 0.48001574 \end{pmatrix} \cdot ohm$$

The rows in the matrix represent the number of electrodes j, the columns represents the influence term per i electrodes influencing j electrodes.

The total influence on electrodes j due to all the i electrodes can be obtained as follow:

$$Gi_k = \frac{1}{N} \cdot \sum_{j=1}^{N-1} \left(Gat_{k,j}\right) \qquad \text{Equation 5.7}$$

Where k runs from 1 to the number of electrodes N.

Implementing this equation in Mathcad® we obtain:

$$Gi_k = \left| \begin{array}{l} \text{for } k \in 1..N \\ \quad Gi_k \leftarrow \frac{1}{N} \cdot \sum_{j=1}^{N-1} \left(Gat_{k,j}\right) \\ Gi_k \end{array} \right. \qquad \text{Equation 5.8}$$

In the above equation, we divide by N to apply the APM (Average Potential Method). Unlike in the previous cases, e.g., Sunde's, Von Baeckmann's, we do not need to multiply by 2 in Equation 5.8 because the matrix calculates the interference for each and all the electrodes (anodes) at once.

Using the values shown above, the total influence on one electrode due to the other electrodes is:

$$Gi = \begin{pmatrix} 0.34258 \\ 0.37015 \\ 0.39933 \\ 0.4643 \\ 0.46152 \\ 0.33621 \\ 0.24041 \end{pmatrix} \cdot ohm$$

The resistance per each anode *without* the interference components is calculated using equations 5.1 or 5.2:

$$Ga_j = \frac{\rho}{4 \cdot \pi} \cdot \frac{1}{L} \cdot \frac{1}{L} \cdot \int_{-(L+t)}^{-t} \left[\ln \left[\frac{L + t + z + \sqrt{(a)^2 + (z + L + t)^2}}{z + t + \sqrt{(a)^2 + (z + t)^2}} \right] + \ln \left[\frac{z - t + \sqrt{(a)^2 + (z - t)^2}}{z - t - L + \sqrt{(a)^2 + (z - t - L)^2}} \right] \right] dz$$

Then, the total resistance for the groundbed is obtained by:

$$Rgb = \frac{1}{N} \cdot \left[\frac{1}{N} \cdot \left[\sum_{i=1}^{N} (Ga_i) \right] + \left[\sum_{i=1}^{N} (Gi_i) \right] \right] \qquad \text{Equation 5.9}$$

or

$$Resist = \frac{1}{\displaystyle\sum_{n=1}^{N} \left[\frac{1}{\left[\left[\frac{1}{N} \cdot \sum_{i=1}^{N} (Ga_i) + \sum_{i=1}^{N} (Gi_i) \right] \right]} \right]} \qquad \text{Equation 5.10}$$

the results for Equation 5.9 and 5.10 are:

Rgb = 0.8544429ohm

and Resist = 0.8544429ohm, respectively

Equations 5.9 and 5.10, are obtained using the APM and the fact that all the electrodes are connected electrically in parallel.

Now, to compare these results with Sunde's, Von Baeckmann, and our developed equations (Chapters 2 and 4) for the case of vertical electrodes with the same length, spacing and depth, we will consider the example shown above, but with the spacing being the same *S=2 m*, *L=7m* and *t=0 m*, the results are as follow:

With Equation 5.9

Rgb = 1.130992ohm

Now following Sunde's approach [3]:

The equation for a single anode:

$$RL = \frac{1}{2\cdot\pi}\cdot\frac{\rho}{L_1}\cdot\left[\ln\left[\frac{4L_1}{d}\cdot\left[1+\sqrt{1+\left(\frac{d}{4L_1}\right)^2}\right]\right]+\frac{d}{4\cdot L_1}-\sqrt{1+\left(\frac{d}{4L_1}\right)^2}\right]$$

Equation 5.11

The factor of interference between anodes:

$$RLt_m = \frac{1}{2\cdot\pi}\cdot\frac{\rho}{L_1}\cdot\left[\ln\left[\frac{2L_1}{S\cdot m}\cdot\left[1+\sqrt{1+\left(\frac{S\cdot m}{2L_1}\right)^2}\right]\right]+\frac{S\cdot m}{2\cdot L_1}-\sqrt{1+\left(\frac{S\cdot m}{2\cdot L_1}\right)^2}\right]$$

Equation 5.12

Sunde's general equation that includes Equations 5.11 and 5.12 and the application of the APM for multiple electrodes is:

$$Rs = \frac{1}{N}\cdot\left[RL+\frac{2}{N}\cdot\left[\sum_{i=1}^{N-1}\left[(N-i)\cdot RLt_i\right]\right]\right]$$ Equation 5.13

without taking the limit for the case of considerable number of anodes, the resistance value is:

Rs = 1.130992ohm

Using the Von Baeckmann equation for multiple vertical anodes groundbed.

Rb = 1.164517ohm

As can be seen, there are no differences between the original Sunde's equation, and Equation 5.9. The difference with the Von Baeckmann equation is about 3%.

Furthermore, using the developed interference factor calculation matrix, from Equations 5.8, with *S=2 m*, *L=7m* and *t=0 m* is:

$$Gi = \begin{pmatrix} 0.5667 \\ 0.68775 \\ 0.74506 \\ 0.76245 \\ 0.74506 \\ 0.68775 \\ 0.5667 \end{pmatrix} \cdot ohm$$

The interesting factor is that the first and last anode have the *least* interference and the anodes in the center of the groundbed the *most* interference. This is exactly what occurs in installed groundbeds, the anodes in the center suffer from more interference, receive less current and consume at a lower rate. ***Hence, we were able to simulate the end effects that are commonly experienced in an installed groundbed and on individual anodes.***

So, Equations 5.8, 5.9 and 5.10 provide a simulation closer to the reality, regarding groundbed consumption and reproduce more accurately the actual behavior of anodes interference in a groundbed. These set of equations are a generalization of the vertical groundbeds. The same procedure can be applied for the case of horizontal anodes as well. The differences between vertical and horizontal will be on the potential equation, and the integration limits.

Some pipeline operators use the positive connection to the groundbed closer to the anode that is at the center of the groundbed, instead of to one end of the groundbed. This procedure suggests that these operators have been using the concept of the resistance of the center anodes is larger for years, even without having a simulation model that justifies this practice. Now, we have a mathematical model that predicts and quantifies this concept. The intention of connecting the groundbed header cable to the center of the vertical or horizontal groundbed is to overcome the fact that the anodes in the center of the arrangement have more resistance to remote earth than the anodes closer to the ends of the anodes' arrangement.

5.1.3 Current per anode calculation.

Using Equation 5.9, or 5.10 we can obtain the groundbed potential to remote earth, multiplying by the current coming from the rectifier. With the condition that the anodes are electrically in parallel and utilizing Ohm's law, the amount of current per anode can be calculated, and more accurately predict the anode's expected life. We can do as follows with an example:

$Itest = 10amp$

Groundbed potential to remote earth using equation 5.9 for the groundbed resistance to remote earth:

$$Vgb = Itest \times Rgb \qquad \text{Equation 5.14}$$

The resistance per anode is calculated as:

$$Raa_j = Ga_j + N \times Gi_j \qquad \text{Equation 5.15}$$

Where j runs through the number of anodes, N is the total number of anodes, Ga is obtained from equation 5.1 or 5.2, Gi is obtained from equation 5.8. And with the values of the example above:

$$Raa = \begin{pmatrix} 7.122 \\ 7.97 \\ 8.371 \\ 8.493 \\ 8.371 \\ 7.97 \\ 7.122 \end{pmatrix} \cdot \text{ohm}$$

The current per anode is then:

$$Ia_j = \frac{Vgb}{(Raa)_j} \qquad \text{Equation 5.16}$$

The current per anodes is then:

$$Ia = \begin{pmatrix} 1.588 \\ 1.419 \\ 1.351 \\ 1.332 \\ 1.351 \\ 1.419 \\ 1.588 \end{pmatrix} A$$

As expected, the anodes in the center have less current than the anodes at the ends.

It is important to mention that the total current is higher than the current output at the rectifier. In this example, the current output at the rectifier is 10 amperes, the total current calculated if we sum the current per anode is:

$$I := \sum_{n=1}^{rows(Ia)} Ia_n$$

$$I = 10.048\,A$$

In this method the resistance per anode includes the interaction between anodes. The interaction between anodes is calculated using the average potential method. When we calculate the average, we are assuming that all the anodes will have the same interaction, which is not exactly true. The error in this case is only 0.5% because all the anodes have the same dimensions, depth and distance between anodes.

If for example, the anodes will have different characteristics; depth, length and spacing between the anodes, the error in the total current calculation is larger, as shown below:

$$L := \begin{pmatrix} 7 \\ 8 \\ 6 \\ 4 \\ 3 \\ 9 \\ 7 \end{pmatrix} \cdot m \quad t := \begin{pmatrix} 1 \\ 3 \\ 2 \\ 4 \\ 5 \\ 2 \\ 3 \end{pmatrix} \cdot m \quad S := \begin{pmatrix} 3 \\ 4 \\ 6 \\ 2 \\ 5 \\ 7 \end{pmatrix} \cdot m$$

The total current calculation is:

I = 10.457A, the error increased almost 10 times to 4.6 %. This happens due to the dissimilarities between anodes is larger than when they have the same dimensions, depth and spacing between them, increasing the error in the calculation of the average potential.

5.1.4 The potential in the ground due to groundbeds with anodes with different length, depth and spacing for vertical electrodes.

The equation to calculate the potential in the soil to remote earth has two components; the first part is due to the first anode expressed below:

$$Vdl(x,y,z) = \frac{\rho}{4\cdot\pi}\cdot\frac{Ia_1}{L_1}\cdot\left[\ln\left[\frac{L_1 + t_1 + z + \sqrt{x^2 + (y-\xi)^2 + (z + L_1 + t_1)^2}}{z + t_1 + \sqrt{x^2 + (y-\xi)^2 + (z+t_1)^2}}\right] + \ln\left[\frac{z - t_1 + \sqrt{x^2 + (y-\xi)^2 + (z-t_1)^2}}{z - t_1 - L_1 + \sqrt{x^2 + (y-\xi)^2 + (z - t_1 - L_1)^2}}\right]\right]$$

Equation 5.17

Then, the influence of the other anodes:

$$Vdmm1(x,y,z) = \sum_{j=1}^{N-1}\left[\frac{\rho\cdot Ia_{j+1}}{4\cdot\pi}\cdot\frac{1}{L_{j+1}}\cdot\ln\left[\left[\frac{L_{j+1}+t_{j+1}+z+\sqrt{x^2+\left(y-\sum_{n=1}^{j}S_n-\xi\right)^2+\left(z+L_{j+1}+t_{j+1}\right)^2}}{z+t_{j+1}+\sqrt{x^2+\left(y-\sum_{n=1}^{j}S_n-\xi\right)^2+\left(z+t_{j+1}\right)^2}}\right]\right]\right]$$

$$Vdmm2(x,y,z) = \sum_{j=1}^{N-1}\left[\frac{\rho\cdot Ia_{j+1}}{4\cdot\pi}\cdot\frac{1}{L_{j+1}}\cdot\ln\left[\frac{z-t_{j+1}+\sqrt{x^2+\left(y-\sum_{n=1}^{j}S_n-\xi\right)^2+\left(z-t_{j+1}\right)^2}}{z-t_{j+1}-L_{j+1}+\sqrt{x^2+\left(y-\sum_{n=1}^{j}S_n-\xi\right)^2+\left(z-t_{j+1}-L_{j+1}\right)^2}}\right]\right]$$

$$Vdmn(x, y, z) = Vdmm1(x, y, z) + Vdmm2(x, y, z) \qquad \text{Equation 5.18}$$

The influence of all the anodes at the point *(x,y,z)* following the additive effect, will then be adding Equations 5.17 & 5.18:

$$Vt(x, y, z) = Vd1(x, y, z) + Vdmm(x, y, z) \qquad \text{Equation 5.19}$$

The variables were defined before in this chapter:

To be able to plot in 3D the potential profile, we must have fixed values of two of the parameters, in this case it will be the length and the depth, keeping the distance between anodes different. The example described below is a groundbed where the anodes have different length, depth and spacing:

The data used for this simulation are:

ρ=30 ohm-meter
I= 10 ampere
N= is the total amount of anodes. (in this case 7)
ξ= 20 meter (distance from the pipeline to the groundbed)

The length of the anodes, depth and spacing are the same used in the previous paragraph.

The graphic solution potential profile is then:

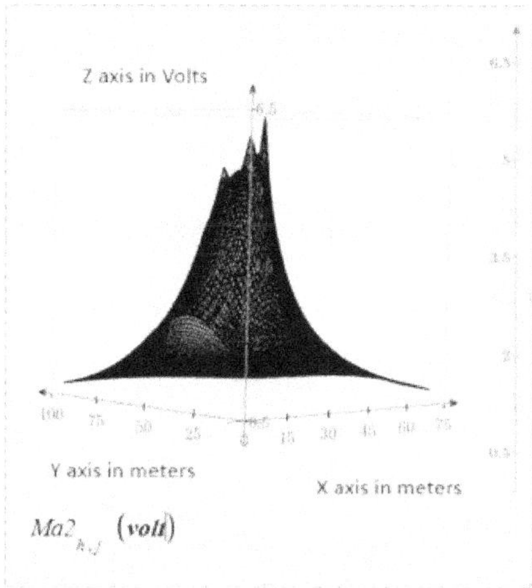

Figure 5-3: Potential to remote earth profile near a vertical groundbed with the length, depth and separation between anodes being different.

The vertical axis provides the potential to remote earth.

In the graph above, the effect of the different anode characteristics is noticeable.

Now, if we use the same equations with same distance, length and depth, using input data L:7m, t:0m and S:10 m, will look like:

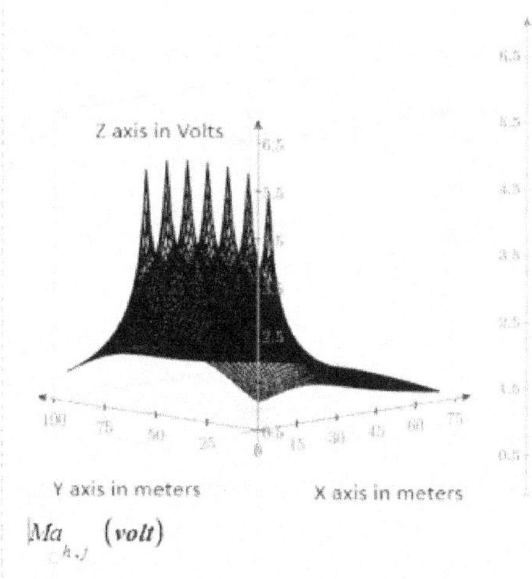

Figure 5-4: Potential to remote earth profile near a vertical groundbed with the length, depth and separation between anodes the same, using the equation 5.19.

5.2 Resistance and Potential calculation when two vertical groundbeds are installed parallel.

5.2.1 Resistance calculation when two vertical groundbeds are installed parallel.

There are some cases where the installation of multiple anodes in a vertical groundbed along a continuous length may be restrictive due to the easement of rights-of-way constraints, Easement restrictions may not provide a sufficient area for placement of all anodes to meet the designed current output, and/or design life, and or meet a targeted groundbed resistance. Yet, the area may be wide enough to facilitate the installation of two vertical shallow groundbeds, paralleling each other. See Figure 5-4 below.

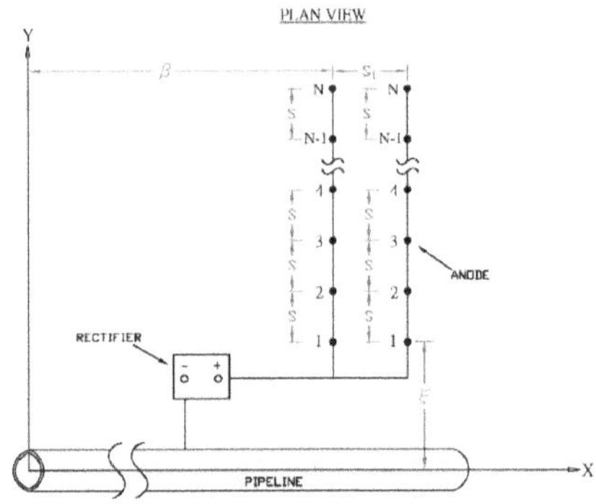

Figure 5-5: Two rows of vertical shallow groundbeds.

Now consider the following:

N: is the number of anodes in each vertical shallow groundbed.
S: is the distance between consecutive anodes in each groundbed and will be the same for both groundbeds. (meters).
S1: is the distance between the two vertical groundbeds. (meters)
β is the distance from y axis to the first groundbed in meters.
ξ is the distance from the x axis to the first groundbed in meters.

See Figure 5-4 above for illustration of the variables defined above.

As discussed in Chapter 4, Section 4.1, the resistance equation for vertical electrodes in a single groundbed has two basic components, as shown in Equations 4.9 and 5.20 below.

$$RL = \frac{1}{N} \cdot \left[\frac{\rho}{2 \cdot \pi} \cdot \frac{1}{L} \cdot \left(\ln\left(8 \cdot \frac{L}{d} \right) - 1 \right) + \frac{2}{N} \cdot \left[\sum_{n=1}^{N-1} \left[(N-n) \cdot R1(n \cdot S) \right] \right] \right]$$

Equation 5.20

The first term is the resistance of an individual vertical anode due to the charges along its axis; the second term is due to the influence of all the other anodes on that one anode.

In the case of two vertical groundbeds connected to the same source and at the distance *S1* between them (Figure 5-4), two interference components should be added to the ones described above:

1. The first deals with the influence of the adjacent anodes in the parallel groundbeds and depends on the value of *S1*.
2. The second is the interference between the remaining anodes and dependent on $\sqrt{(S \cdot i)^2 + S1^2}$, is the distance of all other anodes excluding the one that is direct across from each other. (Figure 5-4)

The final equation will have the form:

$$Rs = \frac{1}{2N} \cdot \left[RL + RLS + \frac{2}{N} \cdot \left[\sum_{i=1}^{N-1} \left[(N-i) \cdot RLt_i \right] \right] + \frac{2}{N} \cdot \left[\sum_{i=1}^{N-1} \left[(N-i) \cdot RLtt_i \right] \right] \right]$$

Equation 5.21

Using the formula from Sunde to calculate each component, they will look like:

- Resistance of the anode depends on the charges on the anode's axis:

$$RL = \frac{1}{2 \cdot \pi} \cdot \frac{\rho}{L} \cdot \left[\ln\left[\frac{4L}{d} \cdot \left[1 + \sqrt{1 + \left(\frac{d}{4L} \right)^2} \right] \right] + \frac{d}{4 \cdot L} - \sqrt{1 + \left(\frac{d}{4L} \right)^2} \right]$$

- Resistance of the anode that is directly in front to one in the parallel groundbed depends on *S1*, the distance between groundbeds.

$$RLS = \frac{1}{2 \cdot \pi} \cdot \frac{\rho}{L} \cdot \left[\ln\left[\frac{2L}{S1} \cdot \left[1 + \sqrt{1 + \left(\frac{S1}{2L} \right)^2} \right] \right] + \left[\frac{S1}{2L} - \sqrt{1 + \left(\frac{S1}{2L} \right)^2} \right] \right]$$

- Resistance term due to the anodes adjacent to other anodes in the same groundbed depends on *Si*.

$$RLt_i = \frac{1}{2 \cdot \pi} \cdot \frac{\rho}{L} \cdot \left[\ln\left[\frac{2L}{S \cdot i} \cdot \left[1 + \sqrt{1 + \left(\frac{S \cdot i}{2L} \right)^2} \right] \right] + \left[\frac{S \cdot i}{2L} - \sqrt{1 + \left(\frac{S \cdot i}{2L} \right)^2} \right] \right]$$

- Resistance term due to the other anodes in the parallel groundbeds

$$RLtt_1 = \frac{1}{2\cdot\pi}\cdot\frac{\rho}{L}\cdot\left[\ln\left[\frac{2L}{\sqrt{(S\cdot i)^2 + S1^2}}\cdot\left[1 + \sqrt{1 + \left[\frac{\sqrt{(S\cdot i)^2 + S1^2}}{2L}\right]^2}\right]\right] + \left[\frac{\sqrt{(S\cdot i)^2 + S1^2}}{2L} - \sqrt{1 + \left[\frac{\sqrt{(S\cdot i)^2 + S1^2}}{2L}\right]^2}\right]\right]$$

The Equation 5.21 can also be applied using Von Baeckmann or the calculations developed in Chapter 2, Section 2.3, Equation 2.7.

We now can derive the degree of influence on the total resistance of the groundbed that is formed by two parallel vertical arrangements, based on the distance between them *(S1)*.

Using the equations above, in this example, and using the following variable values, see Figure 5-4 for the variables' definition:

ρ= 30 ohm-meter
L= 7 meter
t= 0.61 meter
N= is the total amount of anodes. (in this case 7)
S= 1.524 meter
S1= 2.134 meter

Into Equation 5.21, we obtain a total resistance of:

Rs=1.05 ohm

Now, keeping the input variables the same, and only changing the distance *S1* between the two lines of vertical anodes to

S1=15.24 meter

Recalculating *Rs* using Equaton 5.21 we obtain a total resistance of:

Rs=0.767 ohm

As expected, the resistance of the groundbed with two lines of vertical anodes decreases as the spacing between them increases.

5.2.2 Potential calculation when two vertical groundbeds are installed parallel.

The potential equation to determine the influence of a groundbed underground, where $z \neq 0$, that consists of two parallel vertical rows of anodes has two parts:

$$U1(x,y,z) = \sum_{n=0}^{N-1} \left[o \cdot \left[\ln \left[\frac{z+t+L+\sqrt{(x-\beta)^2+(y-n\cdot S-\xi)^2+(z+t+L)^2}}{z+t+\sqrt{(x-\beta)^2+(y-n\cdot S-\xi)^2+(z+t)^2}} \right] + \ln \left[\frac{z-t+\sqrt{(x-\beta)^2+(y-n\cdot S-\xi)^2+(z-t)^2}}{z-t-L+\sqrt{(x-\beta)^2+(y-n\cdot S-\xi)^2+(z-t-L)^2}} \right] \right] \right]$$

Equation 5.22 (Row 1)

$$U2(x,y,z) = \sum_{n=0}^{N-1} \left[o \cdot \left[\ln \left[\frac{z+t+L+\sqrt{(x-S1-\beta)^2+(y-n\cdot S-\xi)^2+(z+t+L)^2}}{z+t+\sqrt{(x-S1-\beta)^2+(y-n\cdot S-\xi)^2+(z+t)^2}} \right] + \ln \left[\frac{z-t+\sqrt{(x-S1-\beta)^2+(y-n\cdot S-\xi)^2+(z-t)^2}}{z-t-L+\sqrt{(x-S1-\beta)^2+(y-n\cdot S-\xi)^2+(z-t-L)^2}} \right] \right] \right]$$

Equation 5.23 (Row 2)

The common term in each of the equations is:

$$o = \frac{I \cdot \rho}{4 \cdot \pi \, 2N \cdot L}$$

Please, note that we divided by 2N to assure that the current from the rectifier Ides is split evenly to all the anodes.

The total potential from the groundbed at the point *(x,y,z)* is:

$$U(x, y, z) = U1(x, y, z) + U2(x, y, z) \qquad \text{Equation 5.24}$$

The variables associated with the equations above have been defined before in this chapter.

The potential profile due to the two parallel vertical anode rows with the values below is shown in Figure 5.5.

ρ=30 ohm-meter
L= 7 meter
t= 0.61 meter
I= 30 ampere
N= is the total amount of anodes. (in this case 7)
ξ= 14 meter
β= 15.24 meter
S=1.524 meter
S1=15.24 meter

These variables were defined in Figure 5-4.

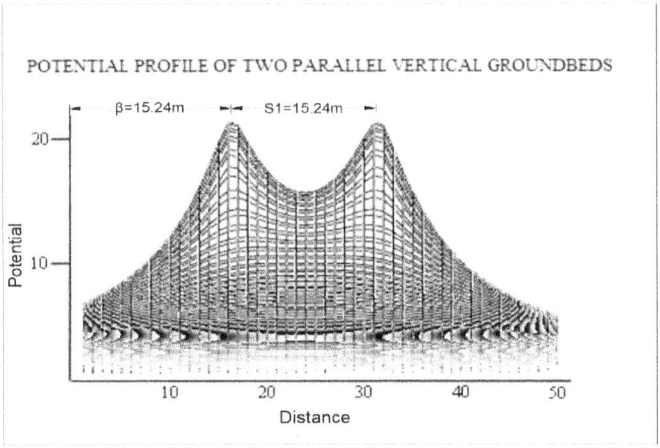

Figure 5-6: Potential to remote earth profile near a vertical groundbed with two parallel rows of anodes.

The vertical axis in Figure 5-5 provides the potential value respect to remote earth in volts.

5.3 Potential calculation when the anodes are distributed non-uniformly.

When the design is at a location like a refinery, a compressor station, or any location where a vertical or horizontal conventional groundbed cannot be easily installed, a distributed anode system may be required.

What conditions should be used to determine the distance between the distributed anodes, in addition to the location of the underground metallic structures?

The objective is to ensure that each underground structure that is cathodically protected receives the influence of the anodes. In this approach, the design strategy is to have the protected structures under the direct influence of the positive potential gradient developed by the anodes in the ground[4].

Hence the procedure is as follows:

1. Calculate the potential profile of an individual anode in the ground to determine the maximum distance between anodes.

Using, as an example for the case of distributed horizontal anodes a variation of Equation 2.10 and Mathcad the maximum distance between anodes can be calculated:

$$
\begin{aligned}
\text{Remdist} = \ & \text{for } y \in 1..1000 \\
& \left| \begin{array}{l}
t1 \leftarrow \dfrac{-t}{ft} \\
x1 \leftarrow \dfrac{L}{2 \cdot ft} \\
z2 \leftarrow \dfrac{z1}{ft} \\
L1 \leftarrow \dfrac{L}{ft} \\
o1 \leftarrow \dfrac{o}{volt} \\
V \leftarrow o1 \cdot \left[\ln\left[\dfrac{x1 + \sqrt{x1^2 + (y)^2 + (t1 + z2)^2}}{x1 - L1 + \sqrt{(x1 - L1)^2 + (y)^2 + (z2 + t1)^2}} \right] + \ln\left[\dfrac{x1 + \sqrt{x1^2 + (y)^2 + (z2 - t1)^2}}{x1 - L1 + \sqrt{(x1 - L1)^2 + (y)^2 + (z2 - t1)^2}} \right] \right] \\
p \leftarrow 1 \\
\text{return} \begin{pmatrix} y \\ V \end{pmatrix} \ \text{if } p \cdot 0.5 \le V \le 0.6
\end{array} \right.
\end{aligned}
$$

The variables have been previously defined. In this example, we specify that the minimum acceptable potential rise in the earth for this design to be between 0.5 and 0.6 Volts. The soil resistivity does influence this limiting value.

The potential in the ground due to one anode can be calculated with any of the equations that were developed in Chapter 2, depending on if the design incorporates horizontal or vertical anodes.

2. Determine the maximum distance between anodes, with the condition that the minimum current density in the ground, around the anode, is at least 0.01 mA/ft^2
   a. The current density is obtained from the gradient in the ground due to an individual anode and the soil resistivity near that anode.
   b. Due to additive effect, several anodes will act at specific points at the same time, the current density at each point will always be greater than 0.01 mA/ft^2
   c. Obviously, the criterion of the current density in the ground due to the anodes influence has not been established in any standard. This value is based on the author's knowledge.
   d. With the procedure developed here, the mathematical tool to apply the condition of current density in the ground will be established.

The equation for the potential gradient calculation is:

$$
gr(x, y, z) = \begin{pmatrix} \dfrac{d}{dx} U(x, y, z) \\ \\ \dfrac{d}{dy} U(x, y, z) \\ \\ \dfrac{d}{dz} U(x, y, z) \end{pmatrix} \qquad \text{Equation 5.25}
$$

Where $U(x,y,z)$ represent the potential in the ground due to one anode, which can be either vertical or horizontal.

The current density is then calculated as follow:

$$\delta I f(x,y,z) = \left| \frac{-1}{\rho} \cdot \sqrt{\left(gr(x,y,z)_1\right)^2 + \left(gr(x,y,z)_2\right)^2 + \left(gr(x,y,z)_3\right)^2} \right|$$

Equation 5.26

3. Assign coordinates to anodes assuring that the distance between anodes complies with the maximum distance determined previously to avoid overprotection of the structures closer to the anodes.

To evaluate this condition Equation 5.38 below takes into account the contribution of multiple anodes influencing the potential on the structure. The anodes coordinates are inserted into a column matrix, the software populates them automatically, and calculates the resulting potential in the ground and these results can be plotted.

$$UT1(x,y,z) = o1 \cdot \left[\ln \left[\frac{x - Xw + \sqrt{(x - Xw)^2 + (y - Yw)^2 + (t + z)^2}}{x - Xw - L + \sqrt{(x - L - Xw)^2 + (y - Yw)^2 + (z + t)^2}} \right] \right]$$

$$UT2(x,y,z) = o1 \cdot \left[\ln \left[\frac{x - Xw + \sqrt{(x - Xw)^2 + (y - Yw)^2 + (z - t)^2}}{x - L - Xw + \sqrt{(x - L - Xw)^2 + (y - Yw)^2 + (z - t)^2}} \right] \right]$$

$$UT(x, y, z) = UT1(x, y, z) + UT2(x, y, z)$$

$$Uts(x,y,z) = \sum_{n=1}^{rows(Xw)} UT(x,y,z)_n$$

Equation 5.27

This equation analyzes the case using multiple distributed horizontal anodes.

Note that $UT(x,y,z)$ has a sub-index to evaluate the potential based on the position of each anode and the column matrices that define the location of each anode as shown below for a practical example:

$$Xw := \begin{pmatrix} 18 \\ 10 \\ -50 \\ 25 \\ 45 \\ 144 \\ 287 \\ 341 \\ 431 \\ 491 \\ 338 \end{pmatrix} \cdot ft \quad Yw := \begin{pmatrix} 225 \\ 306 \\ 345 \\ 407 \\ 342 \\ 358 \\ 429 \\ 473 \\ 470 \\ 368 \\ 368 \end{pmatrix} \cdot ft \qquad \text{Equation 5.28}$$

A 3D plot for this example is shown below:

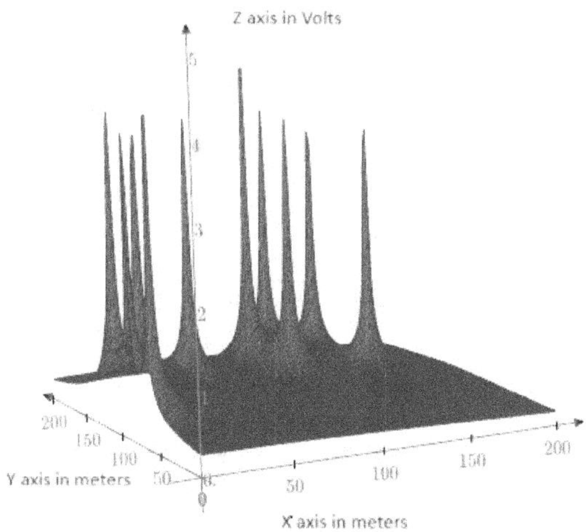

Ms **(*volt*)**

Figure 5-7: Potential profile of a group of anodes distributed in a station.

The vertical axis represents the potential in the ground respect to remote earth in volts. The peaks in the graph coincide with the location of each of the 10 anodes.

It is also possible to plot the groundbed profile in Google Earth and determine how much influence each anode and all the anodes together have in relation to their surroundings. This type of evaluation provides the designer the flexibility to change the location of the anodes before the anodes are installed, to determine the most efficient location to have all the underground structures cathodically protected, minimizing over protection, and shielding issues.

In this example, the resistance can be calculated using any of the equations discussed in Chapter 4 for multiple groundbeds. The average distance between anodes can be obtained using the Mathcad procedure below:

$$\text{Distance} = \left| \begin{array}{l} \text{for } i \in 1..\text{rows}(Xw) - 1 \\ \quad \text{for } j \in i..\text{rows}(Xw) - 1 \\ \qquad D_{i,j} \leftarrow \sqrt{\left(Xw_i - Xw_{j+1}\right)^2 + \left(Yw_i - Yw_{j+1}\right)^2} \\ D \end{array} \right.$$

Equation 5.29

The input values for this subprogram are the coordinates for each anode, as they appear in Equation 5.28 above. As an example, it generates a matrix with the following values, based on *Xw* and *Yw* given by 5.28:

Distance =

	1	2	3	4	5	6	7	8	9	10
1	81.4	137.9	182.1	120.1	183.2	337.6	407.2	480.2	494.1	350.5
2	0	71.6	102.1	50.2	143.7	303.1	370.7	451.8	485	333.8
3	0	0	97.3	95	194.4	347.3	411.4	497	541.5	388.7
4	0	0	0	68	128.7	262.9	322.8	410.9	467.6	315.4
5	0	0	0	0	100.3	257.2	323.7	406.7	446.8	294.2
6	0	0	0	0	0	159.7	228.1	308.1	347.1	194.3
7	0	0	0	0	0	0	69.7	149.7	212.9	79.5
8	0	0	0	0	0	0	0	90	183.1	105
9	0	0	0	0	0	0	0	0	118.3	138
10	0	0	0	0	0	0	0	0	0	153

·ft

The mean value of the matrix above is: mean(Distance) = 140 ft

Because the matrix has $\left(\sum_{i=1}^{Nh-2} i\right)$ of zero components, the mean value calculated above does not represent the average of the distances between all the anodes. To obtain the true average value, we must multiply mean (Distance) by all the total components of the matrix-$(Nh - 1)^2$ and divide by the actual amount of non-zero components.

Amount of non-zero components:

$$(Nh - 1)^2 - \left(\sum_{i=1}^{Nh-2} i \right)$$

The average spacing between anodes is calculated then as follows:

$$S = \frac{\text{mean(Distance)} \cdot (Nh - 1)^2}{(Nh - 1)^2 - \left(\sum_{i=1}^{Nh-2} i \right)} \quad \text{, Equation 5.30}$$

where *Nh* is the total number of anodes. The value of S for this example is 266-ft.

5.4 Special case of APM application. Determination of Crowding factor between vertical and horizontal groundbeds determination.

There are circumstances where there is a mixed of groundbeds with different configurations, where the mutual influence (crowding factor) is not a straightforward application of the average potential method (APM).

In this Section, we are developing an application of the APM in which an existing horizontal groundbed formed by two linear anodes is not performing as expected, and a vertical groundbed is installed in its vicinity to improve the current distribution to gas wells. The two groundbeds are connected electrically in parallel to the same rectifier.

How does the interaction between the two groundbeds increase the total resistance of the cathodic protection system? What is the crowding factor due to this interaction, and how does it increase the parallel resistance of the combined groundbed?

5.4.1 Crowding factor components

First, we will calculate the influence of the vertical groundbed on the horizontal groundbed. The horizontal groundbed is formed by two linear anodes that are installed very close. The vertical groundbed is a shallow groundbed with several anodes in individual coke columns. Figures 5-8 and 5-9 below show the relative locations of the anodes for this example and the variables' identification.

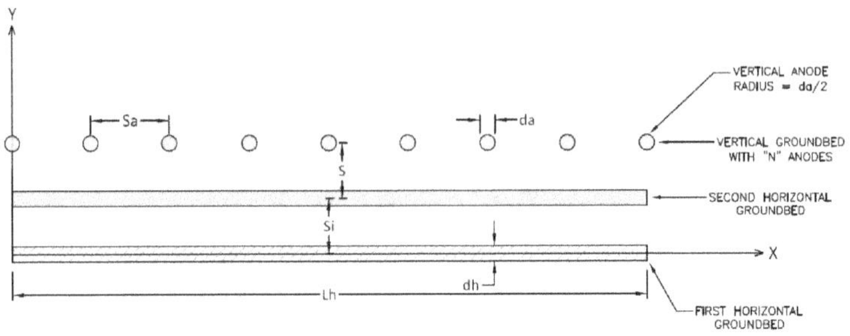

Figure 5-8: Horizontal and vertical groundbeds spatial distribution, plan view.

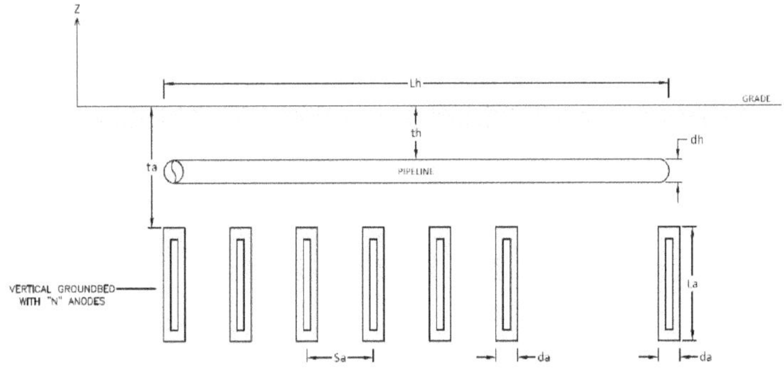

Figure 5-9: Horizontal and vertical groundbeds spatial distribution, lateral view.

The origin of coordinates x and y is arbitrarily selected to be at the beginning of the first horizontal groundbed. All the calculations will be completed with integral equations. As shown before, the limits of integration are one of the most important parts in the application of the concept of the APM. We will explain how they are chosen in each case.

We will obtain all the resistance values using integral equations, where the function to calculate inside the integral is the corresponding potential equation. The resistance components to be calculated are for:

- Each horizontal linear anode and they are compared with Dwight's equation analytical form.
- For the resistance component between the horizontal linear anodes, the crowding factor between the linear anodes.
- For the vertical groundbed.
- For the influence of the vertical groundbed on the horizontal anodes.
- For the influence of the horizontal anodes on the vertical groundbed.

All the calculations will be performed using the values of an actual design to compare the integral values with analytical values. The variables are identified in the figures 5.8 and 5.9, and their values are as follows:

Lh=15.24 meters
La= 4.267 meters
□=147.12 ohm-meter (soil resistivity)
d=0.038 meter
da=0.305 meter
Sa=2.438 meter
Si=0.203 meter
S=1.524 meter
ta=2.134 meter
th= 1.524 meter
Ia=20 ampere (current in the horizontal groundbed)
Ides=15 ampere (total current in the vertical groundbed)
Iaa=current per anode in the vertical groundbed
N=7 (number of anodes in the vertical groundbed)
z=th in the calculations for the horizontal anodes.

5.4.2 Horizontal linear anode resistance and interference between horizontal anodes.

The resistance of the individual horizontal anodes can be obtained using the following integral of the potential in the ground due to the horizontal anode. Notice that:

- The limits of integration go from the beginning to the end of the anode length, running in the x axis.
- The value of the y axis is the radius (half diameter) of the anode.
- Ia, the current is the current of the anode and its image, supposedly concentrated in the axis of the anode and anode's image.
- The division by *Lh* in the integral equation is part of the application of the APM:

$$o := \frac{Ia \cdot \rho}{4 \cdot \pi \cdot Lh}$$

$$RO1 := \frac{o}{Ia \cdot Lh} \cdot \int_0^{Lh} \left[\ln\left(\frac{x + \sqrt{(x)^2 + \left(\frac{d}{2}\right)^2 + (th+z)^2}}{x - Lh + \sqrt{(x-Lh)^2 + \left(\frac{d}{2}\right)^2 + (z+th)^2}} \right) + \ln\left(\frac{x + \sqrt{(x)^2 + \left(\frac{d}{2}\right)^2 + (z-th)^2}}{x - Lh + \sqrt{(x-Lh)^2 + \left(\frac{d}{2}\right)^2 + (z-th)^2}} \right) \right] dx$$

Equation 5.31

The result is: *R01*=12.094 ohm

If, instead of the integral, we use the Dwight equation:

$$R0 := \frac{\rho}{2 \cdot \pi \cdot Lh} \cdot \left(\ln\left(4 \cdot \frac{Lh}{d}\right) - 1 + \ln\left(\frac{Lh + \sqrt{(2 \cdot th)^2 + Lh^2}}{2 \cdot th} \right) + \frac{2 \cdot th}{Lh} - \frac{\sqrt{(2 \cdot th)^2 + (Lh)^2}}{Lh} \right)$$

Equation 2.14

The result is:

R0=12.092 ohm

The difference between the two values is that the Dwight equation above approximates the case when *Lh*>>*th*. So, the integral value is more accurate.

Now, the integral to calculate the interference between the horizontal anodes (mutual effect). The resistance component due to interaction between anodes is obtained using the same integral used before, but with the following characteristics:

- The limits of integration go from the beginning to the end of the anode length, running in the x axis.
- The value of the y axis is the distance between the two horizontal anodes, *Si*.
- *Ia*, is the current, supposedly concentrated in the axis of each anode and their images.
- The equation below is multiplied by two because of the mutual effect and divided by Nh because they are electrically parallel. In this case, Nh=2.
- The division by *Lh* in the integral equation is part of the application of the APM:

$$R1 := \frac{2 \cdot o}{Nh \cdot Ia \cdot Lh} \cdot \int_0^{Lh} \ln\left(\frac{x + \sqrt{(x)^2 + (Si)^2 + (th + z)^2}}{x - Lh + \sqrt{(x - Lh)^2 + (Si)^2 + (z + th)^2}} \right) + \ln\left(\frac{x + \sqrt{(x)^2 + (Si)^2 + (z - th)^2}}{x - Lh + \sqrt{(x - Lh)^2 + (Si)^2 + (z - th)^2}} \right) dx$$

Equation 4.14

R1=8.473 ohm

Explanation of the components in Equation 4.14:

- The horizontal anodes act one on each other, a reciprocal effect considered when we multiply by two in the equation.
- The anodes parallel the x-axis, as shown in Figure 5.8 above; one anode is located on the x-axis for convenience. The second is Si from the first one, which is incorporated in the y value in Equation 4.14.
- The integral is divided by Nh, the total number of anodes, to calculate one anode's average (APM) interference.

Now, the same mutual effect using equations 4.15 and 4.16, in chapter 4:

$$R1s := \frac{\rho}{2 \cdot \pi \cdot Lh} \cdot \frac{2}{Nh} \cdot \left(\sum_{m=1}^{Nh-1} \left((Nh - m) \cdot \left(\ln\left(\frac{Lh + \sqrt{(m \cdot Si)^2 + Lh^2}}{m \cdot Si} \right) + \frac{m \cdot Si}{Lh} - \frac{\sqrt{(m \cdot Si)^2 + (Lh)^2}}{Lh} \right) \right) \right)$$

$R1s$=6.1824 ohm

$$R2s := \frac{\rho}{2 \cdot \pi \cdot Lh} \cdot \frac{2}{Nh} \cdot \sum_{m=1}^{Nh-1} \left((Nh - m) \cdot \left(\ln\left(\frac{Lh + \sqrt{(m \cdot Si)^2 + (2 \cdot th)^2 + Lh^2}}{\sqrt{(m \cdot Si)^2 + (2 \cdot th)^2}} \right) + \frac{\sqrt{(m \cdot Si)^2 + (2 \cdot th)^2}}{Lh} - \frac{\sqrt{(m \cdot Si)^2 + (2 \cdot th)^2 + (Lh)^2}}{Lh} \right) \right)$$

$R2s$=2.2905 ohm

$R12s$=$R1s$+ $R2s$

$R12s$=8.473 ohm

As expected, the integral and analytical solutions for the interference factor obtain the same results.

So, the total resistance of the two horizontal anodes is calculated as follows:

$$Rh := \frac{R01 + R1}{Nh}$$ Equation 5.32.

How does that work?

- This equation adds the resistance of one anode with no interference; R01 above, with the average interference per anode; R1 above.
    - The total resistance of one anode, when influenced by other anodes, is the resistance of the anode to remote earth plus the average crowded factor for one anode.
- Then, it is divided by the total amount of anodes, Nh, because the anodes are electrically in parallel: the resistance of the groundbed is the resistance of one anode divided by the number of anodes.

5.4.3 Interference due to the vertical groundbed on the horizontal anodes.

The current per vertical anode was obtained using the procedure in paragraph 5.1.3 of this chapter.

$$Iaa^{T} = [2.369 \quad 2.126 \quad 2.033 \quad 2.007 \quad 2.033 \quad 2.126 \quad 2.369]\ A$$

The potential factor per vertical anode to be used in the integral is:

$$ov_k := \frac{Iaa_k \cdot \rho}{4 \cdot \pi \cdot La}$$

The average potential factor acting on the horizontal anodes:

$$ovt := \frac{1}{N} \cdot \sum_{n=1}^{N} \frac{ov_n}{Iaa_n \cdot Lh}$$

The interference factor on the first linear anode is calculated using the integral of the vertical groundbed potential integrating on the linear anode location, from *0* to *Lh* in the x-axis, the value on the y-axis is *y1=S+Si*, the distance from the vertical groundbed to the first anode. We then divide by N as in the application of the APM:

$$Rvtt1 := \frac{1}{N} \sum_{p=0}^{N-1} \left[(ovt) \cdot \int_{0ft}^{Lh} \ln\left(\frac{z - ta + \sqrt{(x - p \cdot Sa)^2 + (y1)^2 + (z - ta)^2}}{z - ta - La + \sqrt{(x - p \cdot Sa)^2 + (y1)^2 + (z - ta - La)^2}} \right) + \ln\left(\frac{ta + La + z + \sqrt{(x - p \cdot Sa)^2 + (y1)^2 + (ta + La + z)^2}}{ta + z + \sqrt{(x - p \cdot Sa)^2 + (y1)^2 + (ta + z)^2}} \right) dx \right]$$

Equation 5.32

$Rvtt1 = 3.725$ ***ohm***

The interference factor on the second linear anode is calculated using the integral of the groundbed potential integrating on the linear anode location, from *0ft* to *Lh* in the x-axis, the value on y-axis is *y2=S*. We again divide by N as in the application of the APM:

$$Rvtt2 := \frac{1}{N} \sum_{p=0}^{N-1} \left[(ovt) \cdot \int_{0ft}^{Lh} \ln\left(\frac{z - ta + \sqrt{(x - p \cdot Sa)^2 + (y2)^2 + (z - ta)^2}}{z - ta - La + \sqrt{(x - p \cdot Sa)^2 + (y2)^2 + (z - ta - La)^2}} \right) + \ln\left(\frac{ta + La + z + \sqrt{(x - p \cdot Sa)^2 + (y2)^2 + (ta + La + z)^2}}{ta + z + \sqrt{(x - p \cdot Sa)^2 + (y2)^2 + (ta + z)^2}} \right) dx \right]$$

Equation 5.33

$Rvtt2 = 3.789$ ***ohm***

5.4.4 Interference due to the horizontal linear anodes on the vertical groundbed.

The influence of the first horizontal linear anode on the vertical groundbed is obtained by integrating the potential in the ground due to the first linear anode, but now in the z axis, where the vertical anodes are located, from *-ta-La to -ta*:

$$R12ztt := \frac{\rho}{Ia \cdot La \cdot N} \cdot \sum_{p=0}^{N-1} \left[\int_{-ta-La}^{-ta} \left(\ln\left(\frac{x - p \cdot Sa + \sqrt{(x - p \cdot Sa)^2 + (y1)^2 + (th + z)^2}}{x - p \cdot Sa - Lh + \sqrt{(x - p \cdot Sa - Lh)^2 + (y1)^2 + (z + th)^2}} \right) + \ln\left(\frac{x - p \cdot Sa + \sqrt{(x - p \cdot Sa)^2 + (y1)^2 + (z - th)^2}}{x - p \cdot Sa - Lh + \sqrt{(x - p \cdot Sa - Lh)^2 + (y1)^2 + (z - th)^2}} \right) \right) dz \right]$$

Equation 5.34

$R12ztt = 1.657 \ \boldsymbol{ohm}$

The influence of the second horizontal linear anode on the vertical groundbed is:

$$R22ztt := \frac{\rho}{Ia \cdot La \cdot N} \cdot \sum_{p=0}^{N-1} \left(\int_{-ta-La}^{-ta} \ln\left(\frac{x - p \cdot Sa + \sqrt{(x - p \cdot Sa)^2 + (y2)^2 + (th + z)^2}}{x - p \cdot Sa - Lh + \sqrt{(x - p \cdot Sa - Lh)^2 + (y2)^2 + (z + th)^2}} \right) + \ln\left(\frac{x - p \cdot Sa + \sqrt{(x - p \cdot Sa)^2 + (y2)^2 + (z - th)^2}}{x - p \cdot Sa - Lh + \sqrt{(x - p \cdot Sa - Lh)^2 + (y2)^2 + (z - th)^2}} \right) dz \right)$$

Equation 5.35

$R22ztt = 1.666 \ \boldsymbol{ohm}$

The total resistance of the vertical groundbed was calculated using the equations developed in paragraph 5.1.2 in this chapter.

$Rgt = 5.606 \ \boldsymbol{ohm}$

The total resistance of the vertical groundbed, including the interference due to the two horizontal linear anodes, is:

$$Rvt := Rgt + R12ztt + R22ztt$$

Equation 5.36

The total resistance of the first horizontal linear anode, including the influence of the second linear anode and the vertical groundbed, is:

$$Rh1t := R0 + R1 + Rvtt1$$

Equation 5.37

The total resistance of the second horizontal linear anode, including the influence of the first linear anode and the vertical groundbed, is:

$$Rh2t := R0 + R1 + Rvtt2$$

Equation 5.38

Now, because the two linear anodes and the vertical groundbed are electrically in parallel, the total resistance of the combined system is:

$$Rt := \frac{1}{\frac{1}{Rvt} + \frac{1}{Rh1t} + \frac{1}{Rh2t}}$$

Equation 5.39

$Rt = 5.149 \ \boldsymbol{ohm}$

What would be the resistance if we applied equations 5.9 and 5.10?

$$Rtt := \frac{1}{3} \cdot \left(\frac{1}{3} \cdot (R0 + R0 + Rgt) + \frac{Rit}{3} \right)$$

Equation 5.9

$$Rtt = 5.456 \ \boldsymbol{ohm}$$

$$Rttt := \frac{1}{\dfrac{3}{\dfrac{1}{3} \cdot ((R0 + R0 + Rgt) + Rit)}}$$

Equation 5.10

$$Rttt = 5.456 \ \boldsymbol{ohm}$$

The difference between the results with equation 5.39, when compared with equations 5.9 and 5.10 above, is that there are significant differences between the resistances of the horizontal and the vertical groundbed. So, when calculating the total resistance of groundbeds with different configurations but interacting between them, equation 5.39 is more accurate.

This chapter has shown different groundbed designs and simulation procedures developed for various anode configurations and design needs based on specific field situations. Now that computer calculation programs are so popular, integrals are a very flexible tool for resistance calculation. The integration limits are important in integrals, following the APM principles. The last paragraph helps to show the application of these principles in practical terms.

The significance of this chapter is not just the equations but also conveying that once basic knowledge is understood, the development of practical, real-world applications is endless. Hence, the corrosion engineer can always adjust the basic theory to meet the design requirements of the cathodic protection system.

As a summary of the application of the Average Potential Method to calculate the crowded factor for any configuration groundbeds is shown below:

1. Calculate the resistance of each groundbed to remote earth.
2. Calculate the interaction (crowded factor) of each groundbed to each other.
3. Add the resistance of each groundbed to remote earth to the corresponding crowded factor due to each other groundbed.
4. Calculate the resistance in parallel to all the groundbeds using equation 5.39.
    a. Equations 5.9 and 5.10 can be used when the number of anodes is large and when the anodes have the same resistance to remote earth.

REFERENCES

1. E. D. Sunde, *Earth Conduction Effects in Transmission Systems*, 2nd ed. (New York, NY: Dover Publications, Inc., 1968), p. 73
2. R. De Las Casas, "Methodological Approach to the Average Potential Method to Obtain a Generalized Equation for the Resistance of Vertical Electrodes," CORROSION 2012, paper no. C2012-0001809, Houston, TX: NACE International, 2012
3. E. D. Sunde, *Earth Conduction Effects in Transmission Systems*, 2nd ed. (New York, NY: Dover Publications, Inc., 1968), p. 78
4. A.W. Peabody, R.L. Bianchetti, ed., *Peabody's Control of Pipeline Corrosion*, 2nd ed. (Houston, TX: NACE International, 2000), p. 28-33.

Applications of the Potential Theory from the Underground Pipelines Point of View

The previous chapters focused on developing equations to design anode groundbed systems for the cathodic protection of underground structures.

This chapter will discuss the potential theory's applications to underground pipelines. The developed analyses will interpret how the pipelines respond to the application of cathodic protection.

The theoretical methodology was developed during the first half of the last century. This chapter summarizes the theory and shows its application to practical scenarios. It will also include field measurements to use and interpret the values obtained with the equations shown. The literature mentioned at the end of the chapter provides a detailed development to obtain the equations mentioned in this chapter.

Items discussed will include the following:

6.1. Explaining what is calculated when the attenuation equations are used.
6.2. How to obtain the pipeline potential profile when the groundbed is not remote. The IR (voltage) due to the groundbed(s) will be calculated and coupled with the pipeline attenuation equations.
6.3. What happens when several pipelines with different attenuation characteristics are laid in the same right of way and connected to the same cathodic protection systems (CPS). What should be done to avoid interference or re-circulating currents between the lines.
6.4. How to measure and calculate the attenuation constants used on the attenuation equations for the case of existing pipelines.
6.5. A different approach to applying the DCVG technique to assess the quality of coatings and to categorize the anomalies found during the survey.

6.1 Modeling of pipe-to-soil potential when the groundbed is not remote from the pipeline.

6.1.1 Pipeline attenuation equation.

The differential equation that is solved to obtain the pipeline cathodic protection attenuation behavior is the Poisson differential equation[1]:

$$\frac{d^2}{dx^2}U(x) = \alpha^2 \cdot U(x) \qquad \text{Equation 6.1}$$

This equation describes the pipeline to remote earth potential shift as a function of the distance from the negative connection to the rectifier due to the cathodic protection groundbed. This potential shift is not a polarization potential, but an IR drop measured to remote earth.[2] It can be measured before polarization occurs or after the polarization is already established.[3,4,5]

The groundbed is considered ideally remote from the pipeline; the change in the potential of the soil in the vicinity of the pipeline to remote earth due to the groundbed's presence is not considered in this formulation. Electrically remote also means that the current density received by the pipeline surface will not depend on the distance from the groundbed.

The pipe longitudinal resistance and conductance to remote earth are known and uniform. The analytical solution of this equation for the case when the pipeline is isolated at both ends or when there are consecutive cathodic protection systems with identical characteristics, the changes due to the presence of the CPS can be measured at the end of the protected section: an insulating flange or the middle point between equally spaced CPS.

$$U_{x,n} = -\left(U(0)\cdot\cosh\left(\alpha p_n \cdot x\right) - Rgp_n \cdot Is1_n \cdot \sinh\left(\alpha p_n \cdot x\right)\right) \qquad \text{Equation 6.2}$$ [6,7]

Where:

U and *U(0)* are the difference between the ON potential due to the cathodic protection current flowing and the OFF potential, both measured to remote earth. *U* can be measured at the very moment when the CP system is turned ON, with zero polarization at the pipeline surface or after polarization has been reached.

- *U(0)* is the value where the pipe is attached to the source (rectifier or sacrificial anode). Note: We are not saying the point on the pipeline closest to the groundbed but where the current returns from the pipeline to the rectifier, the negative connection between the pipeline and rectifier.
- *α*p represents the pipeline attenuation constant, which is dimensionless.

- x represents the units of length where the pipe potential to remote earth is being calculated. It is also dimensionless.
- I_s is the current injected at the groundbed and returned by the pipeline to the rectifier in ampere.
- *Rgp* is the pipeline characteristic resistance in ohm.
- $U_{x,n}$ is calculated where the values of x increment moving away from the negative connection on the pipeline.
- n index runs on the number of pipelines in the case of multiple pipelines in the same right-of-way.

If the middle point between two CPS or the location of the insulating flanges in the case of only one CPS is at a distance L from the CPS, Equation 6.2 can also be expressed as:

$$U_{x,n} = \frac{-U(0)\cdot\cosh\left[\alpha p_n\cdot(L-x)\right]}{\cosh\left(\alpha p_n\cdot L\right)}$$ Equation 6.3[8]

Where L is measured in units of length.

Equations 6.2 and 6.3 above are applicable where the current at the middle point between consecutive identical CPS and at the insulating flanges at the end of the protected section is zero.

If the pipeline has only one CPS, the pipeline's end is very far from the CPS, and the presence of the CPS is not measurable at the end of the protected section; hence, the pipe behaves as an infinitely long pipe; the solution of Equation 6.1 is:

$$U_{x,n} = -U(0)\cdot e^{-(x)\cdot\alpha p_n}$$ Equation 6.4

The definitions of constants and variables are the same as before.

Generalizations of Equations 6.2, 6.3, and 6.4 are discussed in paragraph 6.2 below.

6.1.2 The analytical equation to determine the groundbed influence in the soil concerning remote earth.

The differential equation that should be solved to obtain the groundbed potential influence in the ground, measured to remote earth, is the Laplacian equation for the buried electrodes (anodes).

$$\Delta V = 0$$ Equation 6.5

Where:

- Δ is the Laplacian operator
- V is the potential in the ground due to the groundbed, measured to remote earth.

The Laplacian equation can be solved using the following assumptions:

- The anodes have a cylindrical shape, where the length of the anode is much larger than its diameter $L >> d$.
- The images on the ground surface due to the charge electrodes are considered.
- The Average Potential Method is applied.

Chapters 1 through 5 of this book presented the equations for different anode configurations.

Example of these equations for the case of a horizontal groundbed with anodes in a parallel distribution but the groundbed perpendicular to the pipeline is:

$$(Uperp1)_x = o \cdot \sum_{p=0}^{Nh-1} \left[\ln\left[\frac{\sqrt{x^2 + \left(y - p\cdot\frac{Sh}{uL} - \frac{Dist}{uL}\right)^2 + \left(z + \frac{t}{uL}\right)^2} + x}{\sqrt{\left(x - \frac{L}{uL}\right)^2 + \left(y - p\cdot\frac{Sh}{uL} - \frac{Dist}{uL}\right)^2 + \left(z + \frac{t}{uL}\right)^2} + x - \frac{L}{uL}} \right] + \ln\left[\frac{\sqrt{x^2 + \left(y - p\cdot\frac{Sh}{uL} - \frac{Dist}{uL}\right)^2 + \left(z - \frac{t}{uL}\right)^2} + x}{\sqrt{\left(x - \frac{L}{uL}\right)^2 + \left(y - p\cdot\frac{Sh}{uL} - \frac{Dist}{uL}\right)^2 + \left(z - \frac{t}{uL}\right)^2} + x - \frac{L}{uL}} \right] \right]$$

Equation 6.6[9]

- *Uperp1* is the potential in the ground due to the horizontal groundbed in volts.
- *Sh* is the distance between the horizontal anodes in meters.
- *Dist* is the distance (meter), between the first anode and the pipeline. Ideally, this value is determined considering that the groundbed's influence on the protected structure should not be more than 5% of the groundbed's potential for remote earth.[10]
- *L* is the length of each anode in meters.
- *uL* is the unit of length for the attenuation equation, and it is used as a coupling factor between the pipeline attenuation equation and the Laplacian equation solution.
- $o = \frac{Ides \cdot \rho}{4 \cdot \pi \cdot N \cdot L}$ This is a common factor for the potential calculation.
- *N* is the number of anodes
- ρ is the soil resistivity near the groundbed in ohm-meter
- *Ides* is equal to I_s

The equation above uses normalized distances. Normalized distances are the distance from the pipeline connection to the rectifier to the point where the potential is calculated and divided by an arbitrarily chosen normalization factor, a fixed distance *uL*. In this way, the variable x used to calculate the potential to remote earth in the ground is dimensionless and compatible with the potential shift of the pipeline to remote earth.

The variables that provide the distance from the groundbed to the point where the potentials in the ground are calculated, y, and the depth of the pipeline, z, are also normalized. As seen in Equation 6.6, all the variables are divided by uL, the chosen length unit. Based on this coordinate system, we calculate the potential influence from the groundbed on the location where the pipeline is located, so the variable y from Equation 6.6 is zero.

The following examples of potential attenuation graphs will show how the distance between the groundbed and the pipe and the resulting potential shift in the ground influence the pipe to soil potentials.[11] All the potential shift values were calculated as follows:

- Using a form of Equation 6.7, which will be discussed in detail in paragraph 6.2.1.
- All potential values are shown positive to facilitate graphing but are negative numbers.
- For these examples, the pipe-to-soil native potential was given as 0.5 volts to Cu/$CuSO_4$. If actual native potentials are measured, those actual values per unit of length can be introduced in the calculation.
- Consider that the polarization is zero. The attenuation is calculated for the instant when the rectifier is turned ON, and it will represent the On potential values.

In the example shown in Figure 6-1, the groundbed configuration has six horizontal anodes parallel to each other and parallel to the pipelines. The groundbed is perpendicular to the pipelines. The two parallel pipelines, one 40-in and one 20-in diameter, are shown with the groundbed located at 51 m and 41 m, respectively. Those distances reflect remote earth, as calculated using the procedure described in Chapter 3, Section 3.1.

The calculated influence of the groundbed on each pipeline using Equation 6.7 is shown graphically in Figure 6-2.

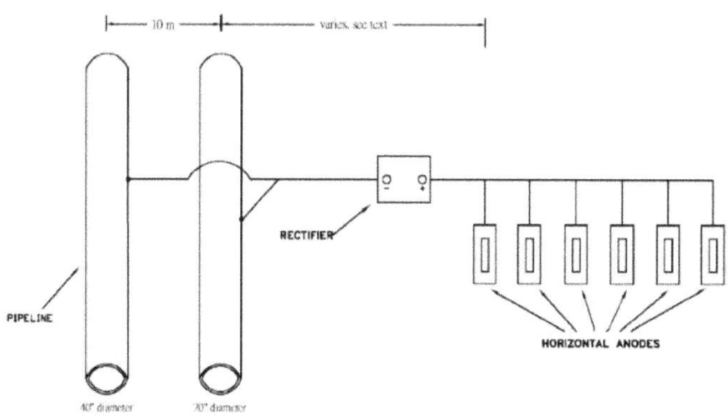

Figure 6-1: Model layout for the example explained above.

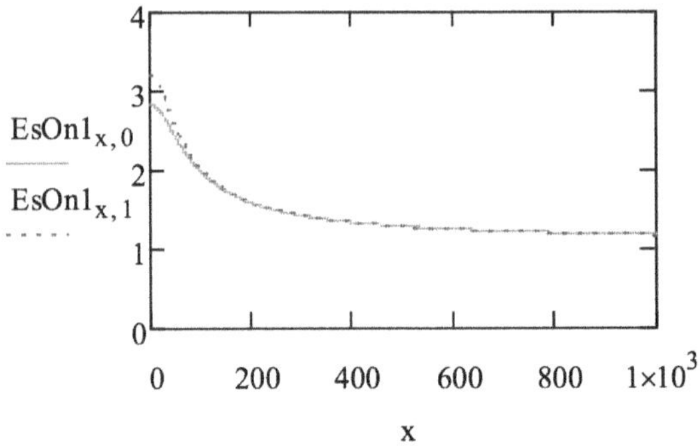

Figure 6-2: The solid red line represents the ON potential to remote earth of the 40-in line, the dotted blue line represents the 20-in line. Groundbed is placed 51 m from the 40" line and 41 m from the 20-in line.

The pipe-to-soil potential (ON) of the 40-in line, which was 51 m from the first anode, is 2.842 Volts to remote earth at the point where the pipeline is connected to the negative of the rectifier.

The 20-in line's pipe-to-soil potential (ON), 41 m from the first anode, is 3.201 volts to remote earth, respectively.

Now, if we move the groundbed closer to the pipes, 25 meters from the 40-inch pipeline and 15 m from the 20-inch pipeline, the new data obtained with the simulation is in Figure 6-3:

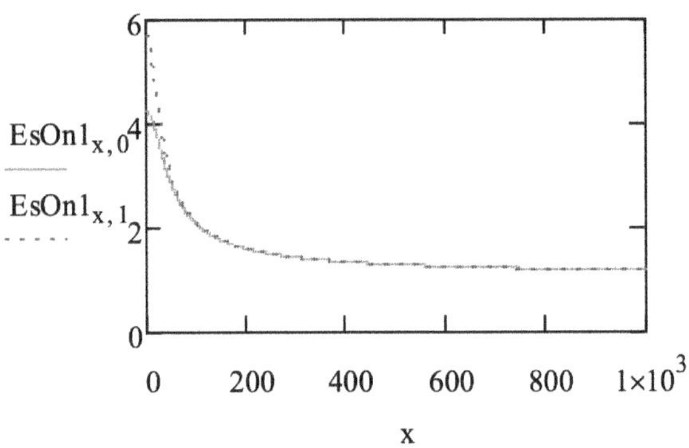

Figure 6-3: The solid red line represents the ON potential to remote earth of the 40-in line, the dotted blue line represents the 20-in line. Groundbed 25 m from the 40" line and 15 m from the 20-in line.

The 40-inch pipeline's potential at the closest point is 4.247 volts, and the 20-inch pipeline's potential at the closest point (10 m closer) is 5.703 volts.

An increase in potential close to the groundbed is observed in both lines.

Now, if we move the groundbed in the opposite direction, let's say 200 meters from the pipelines, the influence on the pipeline potentials to remote earth is as follows. The 40-inch pipeline potential at the closest point is 1.593 volts, and the 20-inch pipeline potential at the closest point (10-m closer) is 1.618 volts (Figure 6-4).

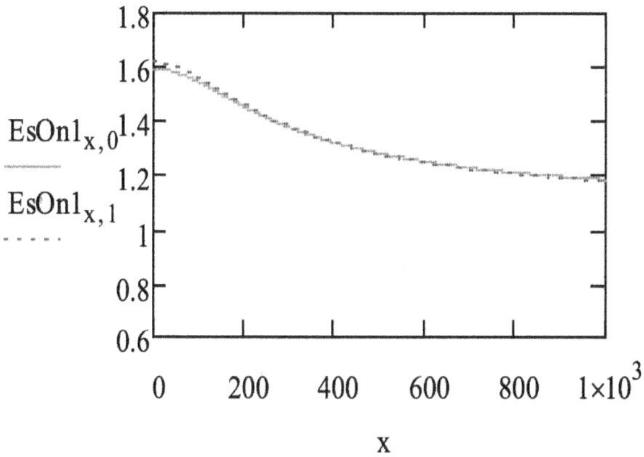

Figure 6-4: The solid red line represents the ON potential to remote earth of the 40-in line, the dotted blue line represents the ON potential for the 20-in line. Groundbed 200 m from the 40" line and 190 m from the 20-in line.

The decrease in the groundbed influence (IR drop added due to the groundbed placement), from 51 m to 200 m changed by 43.9%.

Now, if we move the groundbed, let say 500 meters, Figure 6-5, the influence is as follow:

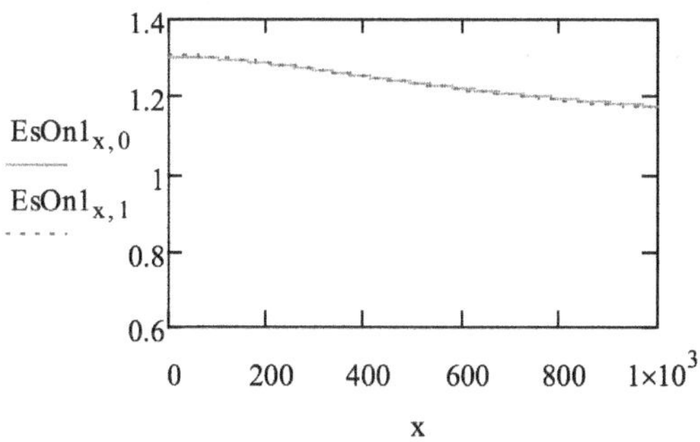

Figure 6-5: The solid red line represents the ON potential to remote earth of the 40-in line, the dotted blue line represents the ON potential for the 20-in line. Groundbed 500 m from the 40" line and 490 m from the 20-in line.

Now the 40-in pipeline potential at the closest point is 1.302 volt and the 20-in pipeline potential at the closest point (10-m closer) is 1.306 volt. Note that the potential difference between the two pipelines is negligible.

The decrease in the groundbed influence (IR drop added due to the groundbed placement), from 200 m to 500 m, changed by only 18.3 %.

As can be seen, moving the groundbed farther away from the pipeline does not significantly decrease its influence on the pipelines; hence, the designer will not gain that much if the groundbed is moved from 200 m to 500 m, in comparison to *moving from 51 m to 200 m*.

Now, if we consider the attenuation calculation with no groundbed IR influence, the pipe-to-soil potentials (ON to remote earth) at the connection point to the rectifier are 1.1 volts for the 40-inch pipeline and 1.1 volts for the 20-inch pipeline. Because the groundbed IR is not considered, the distance between the pipelines and the groundbed is infinitely remote.

The simulation tools developed here can represent any groundbed configuration for pipelines and/or several pipelines with different attenuation characteristics, considering the influence of the groundbed as well, as discussed in the following paragraph.

These calculations also allow the designer to determine the optimum distance to locate the groundbed from the structure to be protected properly and any potential foreign line structure (interference) before installing the groundbed, saving money and time for the pipeline owner and engineering company.

6.2 Modeling a multiple pipeline corridor with a shared cathodic protection system (CPS).

Transmission pipelines sharing a common right of way (ROW) and cathodic protection system (CPS), having different diameters, coating characteristics, and ages (different attenuation characteristics), will benefit from the CPS differently. When the CPS is turned off during regular maintenance, annual or CIS surveys, recirculating currents appear in less than a second. These recirculating currents will move from the pipelines with a smaller attenuation constant (well-coated pipelines) to the lines with a larger attenuation constant (poorly coated pipelines), through the ground, and return through mutual bonds and/or negative junction boxes to the pipeline with smaller attenuation constant. This phenomenon can initiate corrosion in the pipelines with a lesser attenuation constant, even with the CPS rectifiers ON. Previous authors have studied these issues, where extensive data have been analyzed and interpreted.[12]

This section offers:

1. A mathematical model to determine how the pipeline attenuation characteristics affect a pipeline's potential, with no interaction from other lines in the ROW

2. The maximum distance between paralleling bonds to avoid recirculating current formation between lines in the same ROW.
3. Explain the interruption method to facilitate performing annual and CIS surveys in pipelines that share a ROW and CPS.

6.2.1 THEORETICAL PROCEDURE

First, we are generalizing equations 6.2, 6.3, and 6.4.

The following equation explains the cathodic protection behavior of underground transmission lines when the pipelines are electrically short and the consecutive CPS are identical:

To one side of the rectifier negative connection:

$$En1_{y,n} = -\left[E0_n \cdot \cosh\left[\alpha p_n \cdot (y - \xi n1)\right] - Rgp_n \cdot In1_n \cdot \sinh\left[-\alpha p_n \cdot (y - \xi n1)\right]\right]$$

Equation 6.7.1

The other side of the negative connection is represented by the following equation.

$$Es1_{x,n} := -\left[E0_n \cdot \cosh\left[\alpha p_n \cdot (x - \xi n1)\right] - Rgp_n \cdot Is1_n \cdot \sinh\left[\alpha p_n \cdot (x - \xi n1)\right]\right]$$

Equation 6.7.2

Now, considering Equations 6.7.1 and 6.7.2 together we obtain

$$Ens1_{\phi,n} := \begin{cases} En1_{\phi,n} & \text{if } 0 \le \phi < \xi n1 \\ Es1_{\phi,n} & \text{if } \xi n1 \le \phi \le \xi s1 \end{cases}$$

Equation 6.7

Where:

En1, Es1, and E_0 represent the difference between the ON potential and the OFF (IR drop) measured to remote earth in the case that the pipeline is already cathodically polarized. It could also be considered the IR shift at the moment when the rectifier is turned on, with no polarization present.

E_0 is the potential value where the pipe is attached to the source (rectifier or sacrificial anode). E_0, *En1,* and *Es1* do not include the groundbed influence at the pipe location where the potential is measured. When the groundbed influence is considered, En1 becomes the true ON potential to remote earth[13], as described in this chapter's previous section.

- *ξn1*: stationing number where the pipeline connects to the rectifier (sacrificial anode system), adjusted by length units.
- *n*: number of pipelines sharing the same ROW and same source(s) of cathodic protection.

- $In1_n$: the current going to each line.
- y: the length units to calculate the pipe potential to remote earth.
- ap_n: attenuation constant per line. The attenuation constant determines how fast $En1_{y,n}$ falls as a function of the distance from the connection point to the power source (in length units). The equation to calculate the attenuation constant is as follows:

$$\alpha p_n = \sqrt{rp_n \cdot gp_n} \qquad \text{Equation 6.8}$$

- rp_n: resistance per unit of length of the pipelines; it depends on the steel resistivity and line metallic wall thickness.
- gp_n: leakage conductance per unit of length per line; it depends on the pipeline's external coating electrical characteristics relative to remote earth and soil resistivity around the pipeline.
- Rgp_n: pipelines' characteristic resistance. (ohm).

Equation 6.7 is a generalization of Equation 6.2. Equation 6.7 facilitates including the connection from the pipeline to several power sources, rectifiers or sacrificial anodes. Multiple connections can be added in the same simulation process, and all the CPSs' influence is calculated on the transmission line.

Equation 6.7 considers that the pipeline(s) are ideally remote from the groundbed. Therefore, Equation 6.7 should be coupled with the groundbed potential profile to account for the IR generated by the groundbed influencing the pipeline(s), as described in the previous section. This analytical study considers the coupling effect of the pipeline potential equation with the groundbed potential equation for the pipelines in the simulation calculations.

The equation then takes the form:

$$EnOn1_{y,n} = EN_{y,n} + Ens1_{y,n} + Uperp1_{y,n} \qquad \text{Equation 6.8}$$

- $EnOn1_{y,n}$ represents the ON potential to remote earth of each pipeline in the right of way connected to the same rectifier.
- $EN_{y,n}$ pipeline's native potential.
- $Ens1_{y,n}$ (Equation 6.7), above.
- $Uperp1_{y,n}$ groundbed's influence in the ground measured respect to remote earth. Its shape will depend on the groundbed characteristics.

Note: The polarization component is not considered in the equation above. The equation above only includes native potentials and IR components to remote earth. Predictions of polarization potentials are not within the scope of this chapter.

If the pipeline(s) are influenced by other rectifiers or bonds with foreign lines, the IR due to these other sources can also be added to Equation 6.8.

The following examples show attenuation curves of existing pipelines in the same ROW, as calculated utilizing the equations above and pipeline attenuation variables obtained during actual field surveys. It can be noticed that, even when the rectifiers might have similar current and voltage output, but due to the distance between the groundbeds and the pipe are not the same, the total IR, considering the groundbed distance to the pipeline, are not the same.

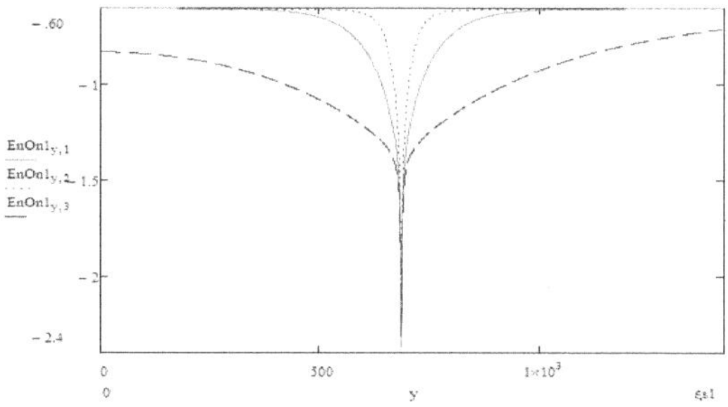

Figure 6-6: Attenuation curves for two existing lines: 12" red & 8" blue and a theoretical new 12" (dotted black pipeline), with only one CPS

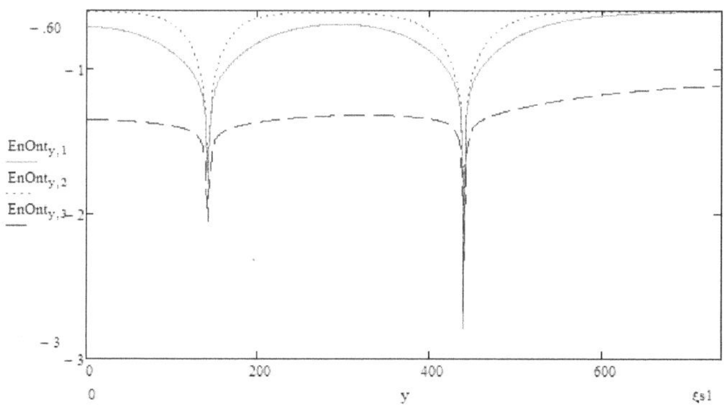

Figure 6-7: Attenuation curves for two existing lines, 12" & 8", and a hypothetical new 12" pipeline, with two CPS

The graphs shown above cannot be obtained by direct measurements in the field due to the circulating currents between adjacent pipes. Pipes connected to a shared rectifier's negative terminal behave like resistances connected electrically in parallel. Since the pipelines are long conductors, they do not perform like pure electrical resistors but like resistors in parallel with specific attenuation characteristics. These attenuation characteristics accentuate as the potential readings are taken further from the rectifier's negative connection. Curves like the ones shown in Figures 6-6 and 6-7 can be obtained from shift (IR) potentials measured to remote earth, assuming the interaction between lines is not significantly affecting the pipe-to-soil readings, i.e., the circulating current between lines is insignificant and is not being considered.

Some operators opt for coupon test stations, where the reference electrodes are incorporated into the coupon's housing, to avoid those circulating currents between lines that may affect the readings. In this way, the reading of the IR drops due to stray/circulating currents with the CPS current OFF on the interruption cycle is minimized.

The authors also developed a conditional program that compares the pipe-to-soil potential for each line in the ROW, considering the pipeline stationing number. The program provides the corresponding stationing number when the potential difference between lines reaches a predetermined value. From there, the location of bonds can be determined to avoid stray/circulating currents between lines.

The conditional program will look like Equation 6.9 below if written in Mathcad®.

$$\text{Lengthln} = \left| \begin{array}{l} \text{for } k \in \xi n1, \xi n1 - 1 .. 1 \\ \quad \left| \begin{array}{l} \Delta En \leftarrow \left| En1_{k,1} - En1_{k,2} \right| \\ \text{return} \begin{bmatrix} \xi n1 - k \\ (\xi n1 - k)\cdot\frac{uL}{ft} \\ \frac{\Delta En}{volt} \end{bmatrix} \text{ if } \frac{\Delta En}{volt} \geq 0.20 \\ \begin{pmatrix} 0 \\ 1000000 \\ 0 \end{pmatrix} \text{ otherwise} \end{array} \right. \end{array} \right. \qquad \text{Equation 6.9}$$

- k represents the distance from the rectifier connection in units of length.
- uL is the unit of length selected for this simulation, in this case, 500 ft. We are calculating the potential for every 500 ft moving away from the rectifier negative connection.
- ΔEn represents the potential difference between parallel lines.

A 0.20 V (200 mV) value for ΔEn was chosen as the upper limit. This establishes that a tentative bond location is identified whenever the potential difference between consecutive lines equals or exceeds 200 mV. This was selected arbitrarily since no documentation the authors know has identified this value. A value of 200 mV was chosen to show how the methodology could be applied.

Incorporating these fixed inputs above, an evaluation can be performed to determine the distance needed for bond placement. That calculated value will depend on the distance between lines, soil resistivities, cathodic protection current density, and coating condition.

Values calculated for the cases of the pipelines shown above in Figures 6-6 and 6-7

$$\text{Lengthln} = \begin{pmatrix} 24 \\ 1.2 \times 10^{4} \\ 0.2034 \end{pmatrix}$$

The maximum distance to install bonds is, in this case:

$mL1_1 := Lengthln_2 \cdot ft$

$mL1_1 = 2.273\,m$

6.2.2 Attenuation equations for the CPS with different characteristics.

If, as often occurs, the consecutive CPSs in a transmission line are not identical, a combination of Equation 6.4[14][15] should be used.

The application of an equation of this type is shown with an example:

A pipeline running west to east has the design calculations for two consecutive CP systems far from the pipeline ends. The protected section has different attenuation characteristics, which prevent using Equations 6.7.

The equations that represent the first rectifier system to the west are:

$$Ew1(x11) := -E0 \cdot e^{-(\xi n1 - x11) \cdot apw1}$$

Equation 6.10.1 represents the potential calculated to the west of the westernmost rectifier system.

$$Ee1(x12) := -E0 \cdot e^{-(x12 - \xi n1) \cdot ape1}$$

Equation 6.10.2 represents the potential calculated to the east of the westernmost rectifier system.

$$Ewe1(x) := \begin{cases} Ew1(x) & \text{if } \xi 0 \le x < \xi n1 \\ Ee1(x) & \text{if } \xi n1 \le x \le \xi s \end{cases}$$

Equation 6.10 represents the combination of the two sides of the negative connection to the westernmost rectifier.

ξn1: represents the stationing number (mileage) where the westernmost rectifier is located, measured in units of length.

ξ0: represents the stationing number (mileage) of the beginning of the section under study, measured in units of length. It is located at the westernmost end of the section.

ξs: represents the stationing number (mileage) of the end of the section under study, measured in units of length. It is located at the easternmost end of the section.

The values of ξ0 and ξs are selected far enough from the two rectifiers that the influence of the CPS is not noticeable.

A graph representing the values obtained with equation 6.10 is as follows:

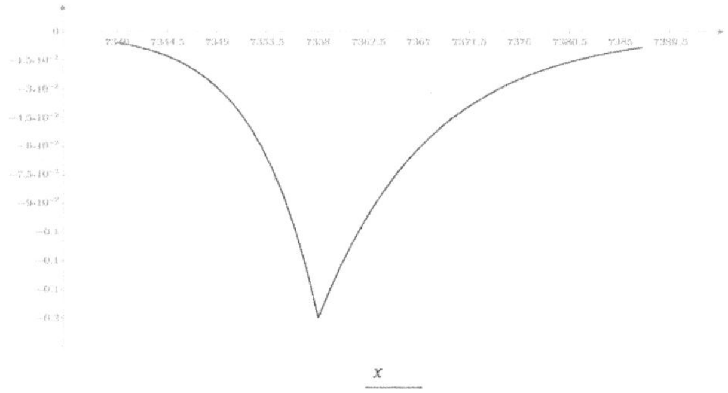

Figure 6-8: Attenuation curves representation of equation 6.10

The equations that represent the second rectifier, the one to the east are:

$$Ew2(x21) := -E0 \cdot e^{-(\xi n2 - x21) \cdot apw2}$$

Equation 6.11.1 represents the potential calculated to the west of the easternmost rectifier system.

$$Ee2(x22) := -E0 \cdot e^{-(x22 - \xi n2) \cdot ape2}$$

Equation 6.11.2 represents the potential calculated to the east of the easternmost rectifier system.

$$Ewe2(x) := \begin{cases} \text{if } \xi 0 \leq x < \xi n2 \\ \quad Ew2(x) \\ \text{if } \xi n2 \leq x \leq \xi s \\ \quad Ee2(x) \end{cases}$$

Equation 6.11 represents the combination of the two sides of the negative connection to the easternmost rectifier.

The variables included in the set of equations 6.11 have the same meaning as before, with the exception of:

ξn2: represents the stationing number (mileage) where the easternmost rectifier is located, measured in units of length.

A graph representation of the Equation 6.11 is as follows:

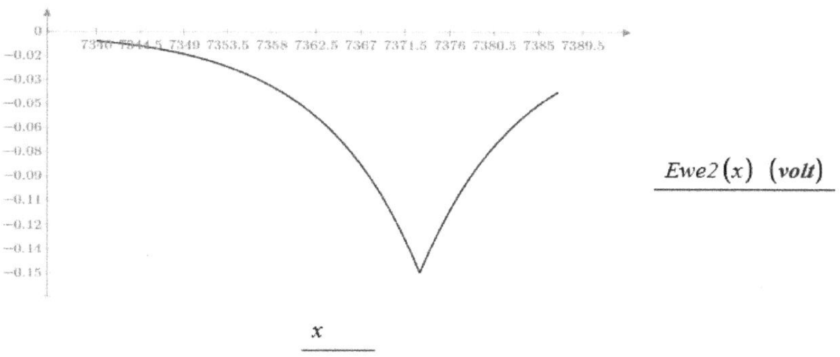

Figure 6-9: Attenuation curves representation of equation 6.11

Now, we can add the two sets of equations as follows:

$$Et(x) := Ewe1(x) + Ewe2(x)$$ Equation 6.12

And a graph representation of the two equations combined is:

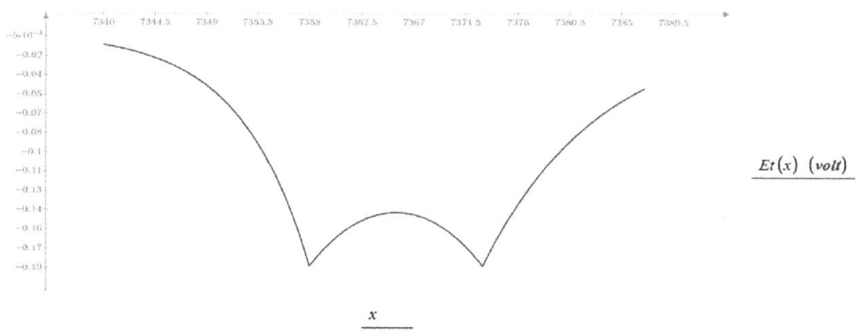

Figure 6-9: Attenuation curves representation of equation 6.12

It is noticeable that the pipeline's attenuation characteristics are not the same at each side of the rectifier stations.

Equation 6.12 can include the native potentials, the rectifier's groundbed influence, and multiple pipelines, like Equation 6.8, as explained in the previous paragraphs.

6.2.3 Experimental procedure

Field data should be gathered to qualify and corroborate the results of the calculations.

When new pipelines have not been installed yet, Equations 6.7 are applicable. The pipeline attenuation characteristics can be chosen from books or any other specialized literature, where the pipeline diameter, wall thickness, and type of coating will be used to estimate the pipeline attenuation constants, with only the soil resistivity, like the field parameter, to be used. In the case of new pipelines with excellent coating, the soil resistivity does not significantly affect the attenuation calculations. The potentials calculated with the attenuation equations will not deviate from the values measured in the field once the pipeline and its CPS are installed.

In the case of existing old pipelines, a different analysis is needed.

The conductance to the ground can be determined from field testing. Several books and standards mention what equations to use, but here, we provide a set of field steps to ensure that the data collected is as accurate as possible.

Field testing steps to determine the attenuation characteristics:

1. If the pipeline has existing CPS near the section under study, ensure that enough of them are off on both sides. This is needed to ensure that a set of equations like 6.10 and 6.11 can be used to calculate the attenuation constants.
    a. The distance selected to determine how many rectifiers should be turned off on each side of the section investigated can be found with a current requirement test at the location where the new CPS will be installed.
    b. If an existing rectifier is at this location, interrupt it and drain as much current as possible from the system. If needed, add a portable groundbed to ensure the current will reach as far as possible on both sides of the rectifier location.
    c. Interrupt this rectifier only and take potential measurements along each test station (TS) on each side of the rectifier to see when the signal from this rectifier is no longer observed, measuring the On and Off potentials, close and remote.
    d. For this testing, the remote potential values, On and Off, are most important.
    e. In the authors' experience, 50 to 60 miles (around 100 km) on each side of the testing section will be enough to keep the existing rectifiers off.
    f. To improve the accuracy of the data to be collected, ensure that foreign rectifiers are not being interrupted and that the bonds with foreign structures are also disconnected.
2. In some instances, existing cathodic protection data like annual surveys and close interval surveys (CIS) can provide enough information to know how far each rectifier is reaching out and, therefore, have those rectifiers **off** for this testing protocol.
3. With all those rectifiers **off**, keep only one rectifier on in the section visited, interrupt it, and take On and Off remote potential values at as many TSs as possible in both directions of the interrupted rectifier.
    a. Ensure that the stationing number of the TS selected for the readings is collected. The pipeline alignment sheets are crucial to knowing the true stationing number.

4. The pipeline owner should be able to provide the pipeline alignment sheets with the "as-built" characteristics so the testing crew would know the pipeline wall thickness, depth, and type of coating that was applied.
   a. Sometimes, the pipeline wall thickness will vary at road, railroad, highway, and river crossings.
   b. Including all those changes in the pipeline wall thickness in the calculations could be cumbersome and, in our experience, unnecessary.
5. Select a unit of length for the following calculations. Once the design is prepared, this unit of length will also be selected for the attenuation calculations.
   a. Any length unit is acceptable.
   b. The engineer in charge of the design should maintain that same unit for all the calculations.
6. This method's limitation is that it only considers the On minus Off to remote earth at TS (potential shift), which hides any local coating damage or possible short to foreign structures that can occur between TS.
   a. That is why it is important to measure the potential shift to remote earth at as many TSs as possible, not just at the TSs that bracket the section under study.
   b. Understanding this aspect will allow the calculation of the worst-case scenario to determine the amount of current per section.
   c. The table below shows the calculation of attenuation constants with values obtained for different test station combinations.
      i. Notice that the attenuation constants, if only calculated between the very ends of the protected section between MP 3.8 and MP 9.5, will hide the coating damages found between MP 3.8 and MP 5.5, which were found during a CIS performed before the CPS design field testing.
      ii. In the brackets MP 2.8 and MP 3.2, the pipeline is electrically shorted at a pump station located at MP 2.7.

line name	Pipe Dia. (in)	Pipe wall thickness (mils)	MP	Distance between TS (FT)	MP bracket	Calculated Conductance (S)	Calculated Characteristic Resistance (Ω)	Calculated attenuation constant	Current applied during testing (A)	Remarks
A	30	375	2.8							
A	30	375	3.2	2272	2.8-3.2	18.40	0.010	0.18223	9.4	The pipe is shorted to the closest pump station, that is why the conductance is that high in this section.
A	30	375	3.8							
A	30	375	5.5	9045	3.8-5.5	**5.50**	**0.018**	**0.09976**	**9.4**	The section with the lowest potentials in the last CIS, is the one with the largest attenuation, marked in yellow. As soon as we incorporate more pipe to the east, the attenuation characteristics improve
A	30	375	6.4	13791	3.8-6.4	2.503	0.027	0.06721	9.4	
A	30	375	6.9	16394	3.8-6.9	1.507	0.035	0.05214	9.4	
A	30	375	8.0	22392	3.8-8	0.765	0.049	0.03715	9.4	
A	30	375	6.4	16368	6.4-9.5	0.025	0.271	0.00665	9.4	
A	30	375	9.5	30159	3.8-9.5	0.654	0.053	0.03435	9.4	

1) The calculated values here are using the steel resistivity of 0.135 ohm*mm2/m (13.5 microohms*cm)
1.1) Values calculated using 18 microohms*cm are also available.
1.2) The calculations will be matched with the field data to see which value of steel resistivity is the most adequate for these lines
2) The calculated attenuation constants largely improve to the east of MP 5.5
3) The river crossing, with the worst coating conditions is between MP 3.8 and MP 5.5
3.1) This section showed the lowest potential during the last CIS
4) Rectifier interrupted located at MP 3.47

Figure 6-10: Field Data and Calculation Results for attenuation constants.

7. Knowing the pipeline wall thickness, diameter, and metal resistivity in conjunction with the potential to remote earth-measured will be used in calculating the attenuation constants; the equations used by the authors are as follows:

$$r := \frac{sr \cdot uL}{C \cdot T}$$ Equation 6.13.1

$$r1 := \frac{sr \cdot uL}{\frac{\pi}{4} \cdot \left((D)^2 - (D - 2 \cdot T)^2\right)}$$ Equation 6.13.2

Equations 6.13.1 or 6.13.2 can be used for the same calculation.

Where:

r is the lineal metal resistance in ohm.
D: is the pipe outside diameter.
T: is the pipe wall thickness.
sr: is the pipeline metal resistivity. This value could vary in function of the type of steel for example, used to manufacture the pipeline. If this value is not provided, the authors use the two following values provided in the literature for steel pipelines:

$$sr := \begin{bmatrix} 0.135 \\ 0.18 \end{bmatrix} \cdot ohm \cdot \frac{mm^2}{m}, \quad sr = \begin{bmatrix} 13.5 \\ 18 \end{bmatrix} \mu\Omega \cdot cm$$

C: is the pipe perimeter in the function of the pipe outside diameter.

uL: is the unit of length chosen for the calculations.

$$g_i := \frac{\left(\frac{\ln\left(\frac{E1}{E2}\right)}{Lp}\right)^2}{r_i}$$

Equation 6.14.1 for the case of electrically long pipelines.

$$g_i := \frac{\left(\frac{\operatorname{acosh}\left(\frac{E1}{E2}\right)}{Lu}\right)^2}{r_i}$$

Equation 6.14.2 for the case of electrically short pipelines.

Set of equations 6.14 is the pipeline conductance per unit of length in Siemens.

E1: is the potential shift (potential on minus off measured to remote earth) in volts of millivolts at the TS closer to the rectifier that is being interrupted.

E2: is the potential shift (potential on minus off measured to remote earth) in volts of millivolts at the TS further to the rectifier that is being interrupted.

Lp is the distance between the two TS where E1 and E2 were measured in units of length.

ln: is the Napierian logarithm function.

$$\alpha_i := \sqrt{g_i \cdot r_i}$$

Equation 6.15 is the attenuation constant per unit of length; it is unitless.

$$Rc_i := \sqrt{\frac{r_i}{g_i}}$$

Equation 6.16 is the characteristic resistance per unit of length in ohm.

The equations above will provide all attenuation variables needed for the attenuation calculations.

If the pipeline is electrically long, the set of equations 6.10-6.11 applies to the design using the constants determined as explained above.

If the pipeline is electrically short, the set of equations 6.7 applies instead.

6.2.3.1 Field data collection procedure once the CPS is installed and running.

- <u>Rectifier readings to corroborate theoretical calculation.</u>

The CPS characteristics that should be measured are voltage DC ON and OFF (back voltage) between the rectifier's output terminals, current for each pipeline at the negative junction box (ON and OFF), and pipelines' potential close and to remote earth ON and OFF at the test stations in the section under study.

- **The groundbed influences the remote reference electrode location.** Since the groundbed impacts the ground in its vicinity, it also influences the ground where the reference cell is located. The calculations shown in paragraph 6.2.1 that arrived at Equation 6.8 are useful.
- **The amount of current circulating on each of the pipelines.** If the CPS protects multiple pipelines, the current per line is the difference between the current ON and the instant current OFF measured at the negative junction box. The OFF currents are the stray/circulating currents between lines due to the differences in the pipelines' attenuation characteristics.
- <u>*Interrupters should also be installed at the negative junction boxes and bonds. It permits one to obtain the true OFF-pipe to soil potential, eliminating the circulating currents in the off part of the cycle.*</u>

- **The pipelines with smaller attenuation constants will have more negative polarization potentials at a greater distance from the rectifier.** The potential values do not decay as fast as those for pipelines with larger attenuation constants.
- As an example, the following values were measured in the field at an existing negative junction box:

Actual DCA per line	DCA at Rectifier	DCV Output Rectifier	Back Voltage	P/S ON (mV)	P/S Off (mV)	IR due to GB (mV)	ΔV (Shift)	Pipe Resistance	Leakage Conductance	Attenuation Constant	Coating Resistance
(Amp)	(Amp)	(Volt)	(Volt)	To remote	To remote	Ref. Elect	mV	Ohm	S/ft2		Ohm*ft2
3.30	5.20	9.57	2.46	-2320	-1347	467	-506	0.1535	1.240.E-05	0.00345	80,645
1.90				-2211	-1367	467	-377	0.1986	8.000.E-06	0.00203	125,000

Figure 6-11: Field Data and Calculation Results

Figure 6-8 shows two components of the difference between P/S ON and P/S OFF:

1. <u>ΔV (Shift)</u> which is the IR due to the resistive factor of the distance between the reference electrode and the closest pipeline coating anomalies, "seen by" the reference electrode, times the current pickup at those anomalies from remote earth. This value can be calculated from the attenuation Equation 6.5.
2. *<u>"IR due to groundbed (GB)"</u>* represents the IR in the ground due to the presence of the cathodic protection groundbed.
    a. When the pipe-to-soil readings are taken close to the groundbed, as shown in this case, the "IR due to the GB" could be very significant if the groundbed is shallow. This IR can be calculated and will depend on the groundbed location, and the location of the reference electrode used to measure the remote pipe-to-soil potential. The equations in Chapter 3 of this book are used for these calculations.
    b. The presence of IR due to the groundbed is observed when a CIS survey is performed on underground transmission lines protected by impressed current systems, and it will be negligible a few hundred feet from the groundbed.
    c. The maximum shift in the ON potential calculated with Equation 6.5, does not necessarily coincide with the maximum shift in the ON potential component due to the presence of the groundbed.
    d. A theoretical analysis of this IR and its interaction with the pipeline attenuation curve appears in the previous section of this chapter.

6.2.4 Analysis of the theoretical versus experimental results

As shown in Figures 6-6 and 6-7 above, the potential differences will facilitate circulating currents between pipelines that share the same ROW. Even when the rectifiers are ON, corrosion may occur on the pipelines with smaller attenuation constants.

<u>This behavior is more accentuated when the rectifier(s) are turned OFF.</u>

Field Measurements have shown:

1. While the rectifiers are ON, the pipes' potentials to remote earth are approximately the same for all pipelines when measured close to the DC power source.
   a. This finding corresponds to the theoretical model, as shown in Figures 6-6 and 6-7, and to the fact that the pipelines show electrical parallelism close to the power source.
2. When the rectifiers are ON, the current circulating on each pipeline through the shunt at the negative junction box is related to the pipeline resistance to remote earth following Ohm's law. The current magnitude is the difference between the current ON and the current OFF.
   a. The attenuation characteristics of each line influence the parallel electrical circuit. When the rectifiers are turned OFF, the circulating currents are measured instantly across the shunts at the negative junction boxes.
3. When the rectifier is turned OFF, there is a stray/circulating current between the pipelines. The current travels from the pipeline with a lower attenuation constant through the soil to the lines with a higher attenuation constant. Then, the current returns to the line with a lower attenuation constant through the buss bar-shunt series combination at the negative junction box or closest bond.
   a. In some instances, current with a magnitude of up to 3 amps was instantly measured, returning through the shunt from the pipelines with a larger attenuation constant to the pipeline with a smaller attenuation constant.
      i. The pipelines with larger attenuation constants are cathodic (less electronegative) than those with lower attenuation constants (more electronegative).
      ii. The theoretical model also predicts this behavior, as shown in Figures 6-6 and 6-7 above.
4. Having currents circulating between pipelines through the negative junction box implies:
   a. While performing ON-OFF CIS or annual surveys on pipelines that share a ROW and CPS, **the true pipe-to-soil OFF potential** cannot be measured. Residual current during the OFF part of the cycle prevents the readings of true polarization potentials.
   b. In annual surveys, where pipe-to-soil potentials are measured at the test stations, bond stations, and above-ground appurtenances, the installation of coupons with stationary reference electrodes in the same coupon's housing is a solution to obtain **true OFF potential readings**.
   c. Since the circulating currents can play an important role in data interpretation, in the case of CIS surveys, the rectifier, the connection at the negative junction box, and the bonds need to be interrupted synchronously to obtain **true OFF potential readings**.
      i. It is understandable that in cases where many bonds need to be interrupted, some bonds stay connected during the survey. However, it is important that the surveyor considers the possibility of the circulating current and relies on taking On and Off readings at coupons that can be easily disconnected from the pipelines to measure the true-off potentials as validation of the CIS data.
   d. Additionally, if DCVG or ACVG surveys are performed, the pipeline should be isolated from the rest of the pipelines in the ROW to avoid errors in data collection.

5. The criterion to determine the maximum potential difference expected for pipelines in the same ROW, sharing the CPS, can be obtained.
   a. The simulation tool developed using Equation 6.9 provides potential differences between lines as a function of the distance from the closest source of DC.
   b. It also determines where interference bonds should be installed to avoid corrosion between adjacent pipelines.

6.3 Calculating the coating anomaly characteristics for underground pipelines.

Due to the advances in the technology tools for pipeline indirect inspection methods like DCVG and CIS performed simultaneously, along with the application of potential theory substantially improve the identification and categorization of coating anomalies.

In this section, a procedure is presented to calculate coating anomaly diameter, resistance, current entering (leaving) the anomaly and the current density. The focus will be on the DCVG methodology, and how to improve the categorization of the anomalies that are identified using DCVG.[16]

This procedure measures the potential difference between two electrodes in the ground to locate coating anomalies in underground coated pipelines. The voltage gradient *is not* calculated as part of this technique. The potential differences exist between the two electrodes because they are positioned on the earth's surface, and the potential difference between the equipotential curves, which are generated around the anomaly, is measured.

The DCVG technique does not consider the location of the anomaly on the pipe surface, the pipe diameter nor the pipe depth, to categorize the anomaly.

The potential difference in the ground depends on the current entering the anomaly, the distance from the coating anomaly to the surface of the ground, and the size of this anomaly.

As an example:

A pipeline is 10-ft deep and, due to the presence of an anomaly, the potential difference measured is 10 mV on the surface of the ground, the size of the anomaly cannot be the same as the case when the pipe is only 3-ft deep, and the potential difference is also 10 mV.

Based on the potential theory developed and presented extensively throughout this book, it has been stated that there is a strong influence on the potential measured which is based on the distance from the anomaly. Additionally, the current intensity that arrives at the anomaly has a significant impact on the coating anomaly characterization, and in determining the current density at the coating flaw.

Below we will present a methodology to calculate: the coating anomaly diameter, resistance, current intensity, and current density. These parameters will be calculated using the following variables:

- the potential difference between the two DCVG electrodes,
- the distance from the pipeline to the surface of the ground,
- the soil resistivity near the anomaly,
- the pipe diameter, and
- the pipe ON and OFF potential to remote earth at the anomaly location.

We will also explain the percentage of error introduced in these calculations based on, the pipe diameter, pipeline depth, and the location of the anomaly respect to the pipe axis: whether the anomaly is below the pipe, at the sides or the top of the pipeline.

An example will be used to illustrate the level of error introduced for the case of pipe with small or large diameter, shallow or deep, and coating anomaly location respect to the pipeline axis.

The variables are defined as follow:

ΔUx: the potential difference between the two DCVG electrodes placed at the soil surface, at the anomaly location. One electrode is over the pipeline, and the other one at some distance x perpendicular to the pipeline, remote from the anomaly. This term is known in the DCVG vocabulary as the "Over the Line Remote Earth" potential, and it is identified by the letter OLRE, this value is in volts.

ΔUr: IR drop to remote earth measured at the coating anomaly. This value can be measured during a combined CIS-DCVG survey and will be equal to the ON-OFF pipeline potential to remote earth at the anomaly location. Measured in volts.

$\rho 0$: soil resistivity at the anomaly location in ohm-meter

$\rho 1$: soil resistivity at the surface of the ground in ohm-meter

x: remote earth distance at the anomaly location in meter.

t: pipeline centerline depth in meter.

$r0$: coating anomaly radius in meter.

D: pipeline diameter in meter.

$R0$: coating anomaly resistance to remote earth in ohm.

I: current entering the anomaly in Ampere.

The current at the anomaly is calculated using the following equation:

$$I = \frac{2 \cdot \pi \cdot \Delta Ux}{\rho 0} \cdot \frac{t \cdot \sqrt{t^2 + x^2}}{\sqrt{t^2 + x^2} - t} \qquad \text{Equation 6.9}$$

The resistance of the anomaly is calculated using the following equation:

$$Ro = \frac{\Delta Ur}{I} \qquad \text{Equation 6.10}$$

The anomaly radius is calculated using the following equation:

$$ro = \frac{\pi}{2} \cdot \frac{\Delta Ux}{\Delta Ur} \cdot \frac{t \cdot \sqrt{t^2 + x^2}}{\sqrt{t^2 + x^2} - t} \cdot \frac{\rho 0}{\rho 1} \qquad \text{Equation 6.11}$$

The current density at the anomaly is calculated using the following equation:

$$\delta I = \frac{I}{\pi \cdot ro^2} \qquad \text{Equation 6.12}$$

6.3.1 The error introduced with the above procedure.

Note that the distance from the anomaly to the earth's surface and the anomaly depth cannot be known accurately. Some procedures can be applied with the two electrodes used to find the anomaly that will give you an idea of where the coating defect is, for example, if it is at the bottom of the pipe, the sides or on the top.

Although these procedures are laborious and can take additional time to complete the survey, they will help improve the accuracy of the calculations, as further explained in this section.

If the actual location of the anomaly cannot be identified, the following calculations provide a margin of error based on pipe diameter, depth, and the probable location of the coating anomaly on the pipe surface.

For example, if the pipe diameter is D=6 in and the pipeline center line depth: t=15 ft, then supposed the anomaly is at the top of the pipeline, the anomaly depth will be:

$$t16 = t - \frac{D}{2}$$

The current entering the defect:

$$I16 = \frac{2 \cdot \pi \cdot \Delta Ux}{\rho 0} \cdot \frac{t16\sqrt{t16^2 + x^2}}{\sqrt{t16^2 + x^2} - t16}$$

Percent of deviation as compared with the calculations where the anomaly is presumed to be at the centerline of the pipeline:

$\%Id16 = \left(1 - \frac{I16}{I}\right) \cdot$ 100, the percent is 2.721%

The same percentage of error will apply for the case of determining the anomaly radius.

If the anomaly is at the bottom of the pipeline, the error is 2.77%

If the anomaly is at one side of the pipeline, the error is even lower 0.545%

Now, if the pipe diameter is 20 in, all the other values the same, the percent of error will be:

Anomaly at the top of the pipeline: 8.88%
Anomaly at the bottom of the pipeline: 9.425%
Anomaly at the side of the pipeline: 1.868%

When the pipeline is deep with a small diameter, an error is introduced when calculating the anomaly characteristics: current at the anomaly, diameter, resistance, and current density are proportionally less than in the case of larger-diameter pipelines. This is because what is important is the distance from the anomaly to the surface of the ground.

This procedure will generally assist in anomaly characterization. Individuals performing DCVG surveys and the calculations above should also include in their analysis the level of error introduced in each specific case based on the pipe diameter and depth.

REFERENCES

1 W. Von Baeckmann, W. Schwenk, W. Prinz, eds., *Handbook of Cathodic Corrosion Protection: Theory and Practice of Electrochemical Protection Processes*, 3rd ed. (Houston, TX: Gulf Publishing Company), p. 550.

2. J. Fitzgerald, J. Wagner, W Young, eds., *CP 4-Cathodic Protection Specialist Manual*, 1st ed. (Houston, TX: NACE International, 2000), p. 4:4.

3. J. H. Morgan, *Cathodic Protection,* 2nd ed. (Houston, TX: NACE International, 1987), p 204.

4. W. Von Baeckmann, W. Schwenk, W. Prinz, eds., *Handbook of Cathodic Corrosion Protection: Theory and Practice of Electrochemical Protection Processes*, 3rd ed. (Houston, TX: Gulf Publishing Company), p. 550.

5. W. Von Baeckmann, W. Schwenk, eds., *Handbook of Cathodic Corrosion Protection: Theory and Practice of Electrochemical Protection Processes*, 2nd ed. (Surrey, England: Porticullis Press LTD, 1975), p. 375.

6. J. Fitzgerald, J. Wagner, W Young, eds., *CP 4-Cathodic Protection Specialist Manual*, 1st ed. (Houston, TX: NACE International, 2000), p. 2:42.

7. W. Von Baeckmann, W. Schwenk, W. Prinz, eds., *Handbook of Cathodic Corrosion Protection: Theory and Practice of Electrochemical Protection Processes*, 3rd ed. (Houston, TX: Gulf Publishing Company), p. 551.

8. J. H. Morgan, *Cathodic Protection,* 2nd ed. (Houston, TX: NACE International, 1987), p 202.

9. R. De Las Casas, "New Earth Potential Equations and Applications," Corrosion 2009, paper no. 09543, Houston, TX: NACE International, 2009.

10. J. Fitzgerald, J. Wagner, W Young, eds., *CP 4-Cathodic Protection Specialist Manual*, 1st ed. (Houston, TX: NACE International, 2000), p. 6:45.

11. Robert Pope, "Attenuation of Forced Drainage Effects on Long Uniform Structures," Corrosion, Vol 2, December 1946.

12. L. Krissa, C. Baeté, J. DeWitt, "CP Management of multiples pipeline right of ways," Corrosion 2015, paper no. 5795, Houston, TX: NACE International 2015.

13. R. De Las Casas, "Modelling of Pipe-to-Soil Potential Coupled with Groundbed Potential Profile," Corrosion 2015, paper no. 6345, Houston, TX: NACE International, 2015.

14. J.M. Standring, "Attenuation of Drainage Effects On a Long Uniform Structure With Distributed Drainage," Corrosion, Vol 3, June 1947.

15. Marshall E. Parker & Edward G. Peattie, "Pipe Line Corrosion and Cathodic Protection, 3rd ed. Gulf Professional Publishing, Appendix D "Attenuation Equations", page 160.

16. W. Von Baeckmann, W. Schwenk, eds., "*Handbook of Cathodic Corrosion Protection: Theory and Practice of Electrochemical Protection Processes*," 2nd ed. (Surrey, England: Porticullis Press LTD, 1975), p. 364.

Index

www.ingramcontent.com/pod-product-compliance
Lightning Source LLC
LaVergne TN
LVHW022223200926
848481LV00006B/2101
* 9 7 8 1 5 7 5 9 0 4 9 5 5 *